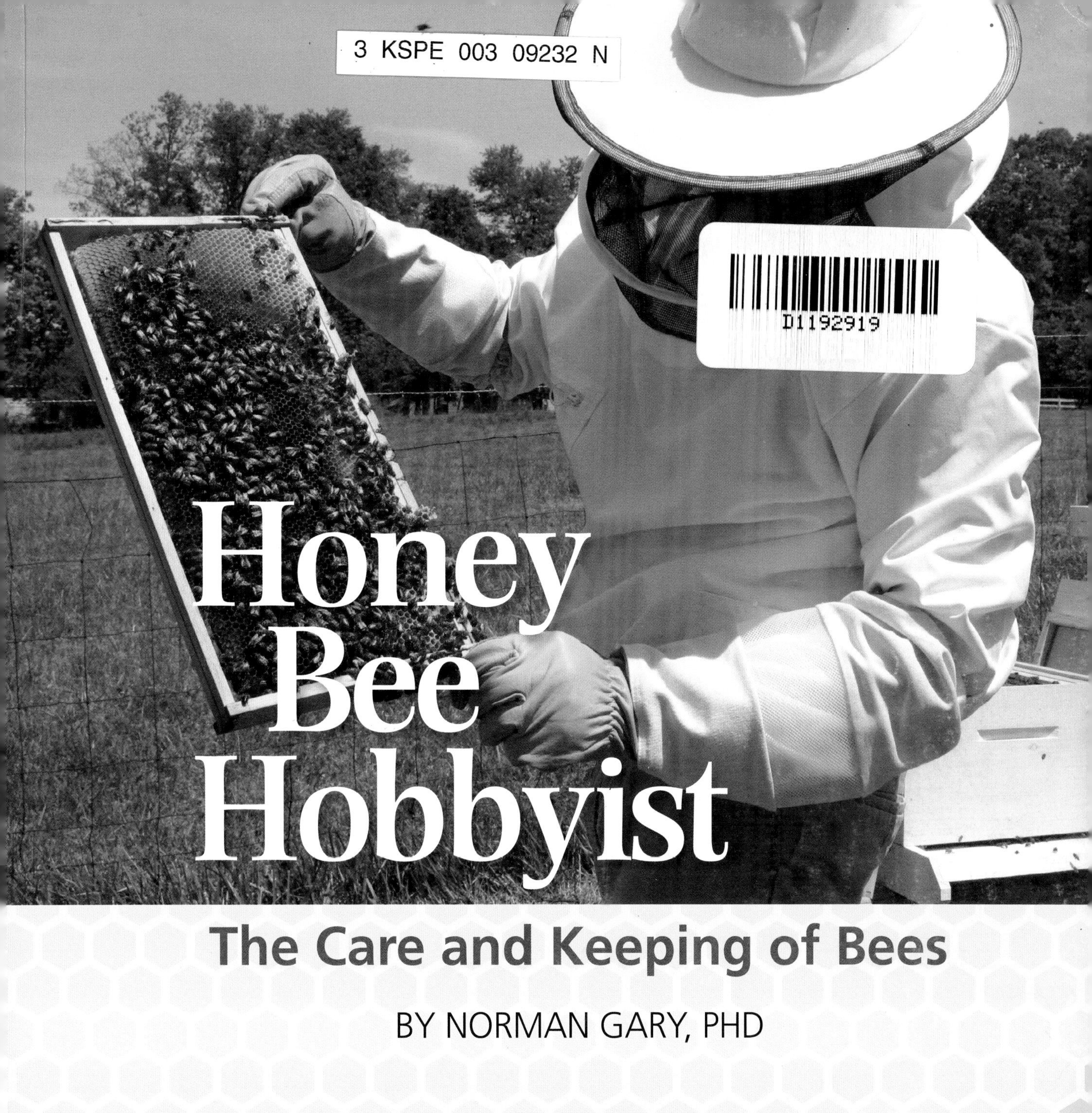

# Honey Bee Hobbyist

## The Care and Keeping of Bees

### BY NORMAN GARY, PHD

HOBBY
H F
FARM
PRESS®

An Imprint of BowTie Press®

A Division of BowTie, Inc.

**June Kikuchi, Andrew DePrisco,** Editorial Directors
**Amy Deputato,** Senior Editor
**Jennifer Calvert, Lindsay Hanks,** Associate Editors
**Elizabeth L. Spurbeck,** Assistant Editor
**Jerome Callens,** Art Director
**Karen Julian,** Publishing Coordinator
**Jessica Jaensch,** Production Supervisor
**Tracy Vogtman,** Production Coordinator
**Melody Englund,** Indexer

Library of Congress Cataloging-in-Publication Data
Gary, N. E. (Norman E.)
Honey bee hobbyist : the care and keeping of bees / by Norman Gary.
   p. cm.
Includes bibliographical references and index.
ISBN 978-1-933958-94-1
1. Bee culture. 2. Honeybee. I. Title. II. Title: Care and keeping of bees.
SF523.G16 2010
638'.1--dc22
             2010020865

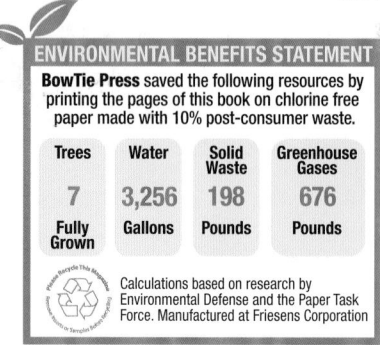

ENVIRONMENTAL BENEFITS STATEMENT
**BowTie Press** saved the following resources by printing the pages of this book on chlorine free paper made with 10% post-consumer waste.

| Trees | Water | Solid Waste | Greenhouse Gases |
|---|---|---|---|
| 7 | 3,256 | 198 | 676 |
| Fully Grown | Gallons | Pounds | Pounds |

Calculations based on research by Environmental Defense and the Paper Task Force. Manufactured at Friesens Corporation

BowTie Press®
A Division of BowTie, Inc.
3 Burroughs
Irvine, California 92618

Printed and bound in Canada
14 13 12 11    2 3 4 5 6 7 8 9 10

# Dedication

To everyone who supported my career with bees: beekeepers, professors, scientists, students, research assistants, movie directors, Hollywood stars, photographers, and family (especially Mom, who never complained about stray bees or tracked honey inside the kitchen)—and to my dog, who led me to the bee tree that started it all...

# Acknowledgments

I am grateful to everyone who contributed in some way in helping me to create this book. Several colleagues in the Entomology Department at the University of California at Davis deserve special recognition: Kathy Keatley Garvey for her wonderful photographic images and for providing a communications link to contributing photographers around the world, Emily Bzdyk for her artistic drawings, Dr. Eric Mussen for sharing his beekeeping expertise, and Susan Cobey for supporting photographic activities at the bee research laboratory on the University campus. Jennifer Calvert, Associate Editor, provided excellent editorial support, especially helping me achieve a tasteful balance between science and art. And special thanks to Sheridan McCarthy who first suggested that I write this book.

# Contents

# Introduction

Honey bees are industrious little animals that astonish newcomers and amaze scientists with their sophisticated social organization and unbelievable communication skills. They do miraculous things. Did you know that one hive can produce 100 pounds or more of honey each year that you can harvest at home? That's especially incredible when you consider that the bees in a hive have to fly more than 50,000 miles to produce a single pound of honey. They routinely fly up to several miles from their hive to collect nectar and pollen from flowers. Bees can even recruit additional help from their sisters, communicating the locations of the flowers by performing a complicated dance—in total darkness—on their honeycomb stage inside the hive. And most importantly, while foraging for their food—nectar and pollen—honey bees also pollinate agricultural crops that account for a third of our food. Pollination is the process in which pollen grains from the male structure of a plant are transferred to the female structure, causing the fertilization that is necessary for the growth of seeds and fruit. Today, as you enjoy consuming a wonderful variety of nutritious, delicious fruits, vegetables, dairy products, cooking oils, and beverages—including your morning cup of coffee—don't forget to thank the bees.

Reading this book is your first step toward a wonderful, exciting, fun, lifelong hobby—beekeeping. Honey bees make great pets (even though actually petting them is a little tricky). They feed themselves. Unlike other pets that require expensive kennel stays when you leave on vacation, you can forget about "bee sitters." Bees don't need daily care and, depending upon the season, can live comfortably for several months without any care at all. They adapt to almost any climate.

Bees are also good neighbors on the farm, even around other animals. Urban environments are ideal for hobby beekeeping because a hive requires so little space; in the city, where backyard space may be scarce or non-existent, some beekeepers locate their hives on rooftops. In addition, honey bees are especially beneficial in urban areas because they pollinate gardens, fruit trees, and other trees and shrubs that provide food (berries and seeds) for wildlife, especially birds.

Beekeeping is less expensive than many hobbies. With proper management, you can even make a profit from the sale of delicious natural honey, fresh from the hive. Imagine opening your hive, removing a natural beeswax honeycomb, and dipping into it with your spoon—or your finger—to sample yummy honey straight from the comb. Fresh, unprocessed honey from the comb is the most delectable reward that nature has to offer.

If you think beekeeping is all work and no play, don't overlook the fun things you can do with bees. Kids love them at first flight. What a great opportunity for them to learn biology in an exciting way. You can even take bees to your child's school and make your son or daughter proud of you for creating the most entertaining and educational presentation of the entire school year. Live bees ("baby" bees that can't sting because they are too young) are very exciting to kids of all grade levels. Imagine the exhilaration—and maybe a few screams—when you release some buzzing drones for free flight in the classroom. (It may be a good idea to first tell the students that drones can't sting because they don't have stingers).

Honey bees, excluding Africanized bees, really are not a significant sting hazard when properly managed and skillfully handled. This book will teach you how to behave when you are around bees so that you can relax and enjoy working and playing with them. It also provides detailed practical instructions for hobby beekeeping, including assembling your new hive, stocking it with bees, properly caring for your bees, and harvesting your honey rewards. Another bonus is that this book contains never-before-published tips that will enable you to become a skillful beekeeper and to avoid bee stings. You can do it. You'll be so pleased to discover how easy it is to start your new beekeeping hobby.

The information in the first chapter is meant to help you decide if hobby beekeeping is for you. Later chapters will discuss in detail what you need to know and acquire in order to fulfill your desire to keep bees.

# CHAPTER 1

# To Beekeep or Not to Beekeep?

Are you a good candidate for hobby beekeeping? Every beekeeper starts out differently, but they all end up pretty much the same—having fun and enjoying an exciting hobby that can last a lifetime. Beekeepers share certain traits. They tend to love animals, enjoy caring for plants, and have gardens. They're also very curious about nature and living organisms in general.

Do you fit this profile? If so, you'll probably love beekeeping. The rewards of having your own hive and harvesting honey—fresh, in various natural flavors, and in quantities that exceed your expectations—make beekeeping an adventure. Beekeeping can also be a profitable enterprise because it's unlikely that you can eat all the honey you produce. You'll be surprised how easy it is to sell your "surplus" honey to neighbors and friends, or at your local farmers' market. Perhaps even more important, yet not so obvious, are the benefits derived from bees pollinating gardens and fruit trees in your backyard and in the surrounding community.

## Bees as Pets

In many respects, beekeeping is easier than caring for conventional pets. A very attractive luxury is that honey bees do not require daily care. You can enjoy caring for your bees when it is convenient for you. They will accommodate your busy schedule. Forget about those time-consuming trips to the pet-supply store for bags of expensive food. Bees gather their own food—nectar and pollen from flowers. And you don't have to be concerned about how much and when to feed them. Bees never eat too much—there are no overweight bees. Under typical circumstances, you don't even have to provide water. And you won't receive expensive bills from the veterinarian. Best of all, bees are quiet—no barking, yowling at night, or crowing at daybreak—and no "poop patrols" to pick up the remains of that pricey pet food.

Bee hives—or "colonies," as beekeepers call them—do require some care. You will need to synchronize colony conditions with seasonal changes and implement the right management procedures at the right time. For example, you have to supply additional combs during periods of abundant nectar secretion to increase the capacity for honey storage. You will also need to protect your colonies from pests, predators, and certain microorganisms.

## The Fear of Stings

Most people have an exaggerated sense of dread concerning bee stings due to a wealth of misleading negative information in the media. With more knowledge and firsthand experience, these fears rapidly vanish. However, such concerns may serve a greater purpose in our society. If not for the fear of stings, too many people might decide to keep bees. Hives in backyards would become commonplace to the extent that there would not be enough nectar and pollen to sustain colonies in overpopulated areas. Bees could not produce harvestable honey under these conditions. They would have to be fed to survive, similar to other pets.

### CAUTION: BEE ALLERGIES

Before you sell your golf clubs to buy beekeeping equipment, you should visit your allergist to determine if you are hypersensitive to honey bee stings. Only about 1 percent of people experience a serious reaction—anaphylaxis. If tests reveal that you are hypersensitive to honey bee venom, stick with golf.

The subject of stings is so important that it is dealt with in great detail in Chapter 8. Sneak a peek now if you just can't wait. After you learn all about stings and have the opportunity to develop hive manipulation skills, you'll see that bees can be controlled easily and are fun animals. You can't teach them to roll over or play dead, but you can train

**You'll be fascinated by the unexpected sounds and sights during your first visit to an apiary. Be sure to see a queen bee and taste some honey right from the comb.**

them to collect artificial nectar (sugar syrup) from a dish on your picnic table. You can even attach a numbered ID tag to each bee so that you can record the number of round trips between their hive and your table. Kids love this experiment, and there's absolutely no sting risk from the foraging bees. (Other interesting activities you can do with bees can be found in Chapter 11.) Honey bees are instinctive critters for the most part. Let them teach you a few tricks, and you and the bees will get along just fine.

## Getting Help from Other Beekeepers

Here's a good strategy for starting your hobby beekeeping adventure: find local beekeepers to help you. Hobby beekeepers are as social as bees are. Most of them would love to share their knowledge. Do an Internet search for "beekeeping supplies" with the name of your city and state. If you are lucky, you'll find a local store that sells beekeeping supplies and may be able to direct you to local beekeepers. Also search for "beekeeping clubs" with your city and state. You're almost certain to find one or more clubs in your area. Attend a beekeeping club meeting and introduce yourself to the club president, who can suggest experienced beekeepers to lead you down the path to success. Approach several club members and tell them about your love of bees and that you need help to get a good start from experts like them (flattery almost always works). Keep "fishing" until you receive one or more invitations to visit a club member's apiary (see Glossary) to see what beekeeping is all about. Other beekeepers should be happy to answer your questions. You can also receive beekeeping information by submitting questions to online beekeeping forums (see Resources).

Beekeeping is not a do-it-yourself operation when you first start. If you want to be a bullfighter, you don't just run out into the ring alone with a red cape. Bees will react instinctively to you, and your reflexive responses will not serve you well. Without thinking, you will do all of the things—like swatting—that cause bees to become defensive and sting. Your adventure would not be fun and could cause you to switch hobbies before you even get started.

When you venture into an apiary for the first time, be sure that your beekeeper host has dressed you in protective clothing in the event that the occasional defensive bee may check you out. You'll need a bee veil over your head and light-color clothing (defensive bees react to dark colors) that covers all exposed skin. Wear long sleeves and full-length trousers. Keep your hands in your pockets so you won't forget and swat at a bee. Now you can relax and enjoy the experience.

You may be nervous—a perfectly normal feeling at this stage of your introduction. Your beekeeping mentor will probably invite you to come close to the hive to get a good view of the bees. As an added safety precaution, politely ask if you can stay about 10–20 feet from the hive that is being opened. The risk of defensive behavior (stinging) diminishes sharply as you move farther away from the hive.

Ask your host to bring combs covered with bees to where you are standing—out of the intense defensive zone near the hive. Ask to see the queen. Which capped comb cells contain honey or a developing brood? What does pollen look like? Where are the eggs? Ask, ask, and ask. If you are not excited at this point, then you should get a pet rat or start growing African violets.

This tagged queen bee isn't wearing a crown and has no makeup. Surrounded by her "court" of worker bees, she lays about 1,200 eggs each day, which are in total nearly equal to her body weight.

After your first positive experience with a living bee colony, you probably won't be able to wait to launch your hobby beekeeping project—but not so fast. First, you need to learn the intimate details of the bees' lives to understand the dynamics of the bee colony. It's important for you to explore the biology and

Bees love lavender. Next time you see some, look for foraging bees, which are interesting to watch, are great for photographers to stalk, and provide entertainment for kids who catch them in cages.

behavior of honey bees. If you don't learn the basics, then you won't experience the full enjoyment of hobby beekeeping. Your bees deserve the best of care. Be sure to read this book—all of it—before you get your first hive. Then read it again after you get your first hive. As you learn the details, you'll enjoy beekeeping more and be able to provide better care for your favorite little critters.

## Where to Keep Your Hives

Where can you locate a hive or two without causing a problem? I hope that you won't have neighbors who have negative feelings about bees. If you do, the best antidote is the promise of free honey and pollination of their fruit trees and gardens. Hives don't require much space, but you do need to leave an open area for flight in front of the entrance. If you're concerned about pets wandering near the hive,

**Nectar and pollen from dandelions stimulate reproduction in bee colonies in the spring.**

you can install a barrier or enclosure to keep them at bay. Such an enclosure can be made using inexpensive materials, such as metal rods (rebar or pipe) pounded into the ground at the four corners and then wrapped with sheer plastic netting—about 1-inch mesh—the kind that is designed to shield plants and trees from deer. In addition to being economical and easy to install, this kind of enclosure is almost invisible. Fortunately, though, pets and most farm animals instinctively tend to stay away from hives. Maybe it's the buzzing sounds that "spook" them.

Don't feel that you have to plant flower gardens to feed your bees. Foraging bees normally range up to 4 miles—that's around 32,000 acres—in all directions to find flowers. Your garden is miniscule by comparison. Just for fun, you could plant a few "bee plants"—plants known to be attractive to bees—so that you can observe their foraging behavior on the flowers.

## When to Start a New Colony

Early spring is the best time to start your new hives. Bees in a new hive may need the entire spring and summer to become established. They have to build combs, rear brood, and accumulate stored honey reserves to survive the winter months when flowers are not available. Starting with a small hive population, around 10,000 bees and a queen, is good because small colonies are usually much less defensive, allowing you to hone your hive-manipulation skills to match the increasing defensive behavior associated with larger colony populations.

It's best to start two hives at the same time—perhaps one full-size hive for honey production and one nucleus hive to maintain a reserve queen (this concept will be explained further in Chapter 9). It's a good idea to get some help from your beekeeping buddies when you order your first packaged bees from a commercial bee supplier.

## Beekeeping Information from the Internet

Can you find all you need to know about hobby beekeeping on the Internet? Absolutely not. A major problem is that there is too much conflicting information, as well as misinformation, out there in cyberspace. Vendors attempt to sell items that you don't really need or can't use, and many amateur beekeepers dole out advice before they are qualified by virtue of experience and knowledge. Hobby beekeepers predictably become totally bewildered by the conflicting information. The total picture of beekeeping knowledge comprises hundreds of bits of information, similar to pieces of a jigsaw puzzle. These bits of information have to be evaluated, integrated, and tested under practical conditions to make sense, and this requires years of experience. In summary, you would have to be an expert, not a beginning hobbyist, at the outset to understand and apply much of the beekeeping information available on the Internet.

Fortunately, you'll be able to find virtually all you need to know about hobby beekeeping in this book. You'll derive the joy and confidence afforded by learning the basics of beekeeping. Even if you already have hives, you'll be excited and pleased to learn new and easier ways to manage your bee colonies.

## USING THE INTERNET IN BEEKEEPING

Continuing education is a vital part of every hobby or occupation. Make it a cornerstone of your beekeeping hobby. Our world is changing—fast—and the world of beekeeping is doing the same. You must strive to obtain the best *current* beekeeping information available to make good decisions about your beekeeping hobby. Fortunately, the Internet is an astoundingly productive source for all information, including beekeeping. But keep in mind that anyone can post information on the Internet, so you must critically analyze everything you read before accepting the information as factual.

The best information is based on good science, so the best sources are institutions that conduct research—universities, state and federal laboratories, legitimate research institutes not operated by commercial interests, and consortiums of authorities from these institutions who are dedicated to discovering the truth and disseminating it. The worst sources for information are sites designed to sell never-before-thought-of miracle products that are touted to do amazing things for your bees. And there are other individuals who simply perpetuate the misinformation they were given. It's unfortunate that many authorities with sterling credentials and a lifetime of practical experience with bees and beekeeping are frequently too busy with professional responsibilities to communicate their knowledge and wisdom to beginning beekeepers. Consequently, amateurs who have the time, energy, and enthusiasm—but not the qualifications—have written many books and articles. Always take information found on the Internet with a grain of salt.

Honey bee pollination produces food for wildlife, especially in urban environments where native pollinating insects are sparse.

## Ordinances that Regulate Beekeeping

If you intend to keep bees in an urban environment, you should find out if there are ordinances that regulate beekeeping operations in your area. Some cities have ordinances that prohibit beekeeping even though it would be in their best interest to permit beekeeping activities to ensure adequate pollination of gardens, fruit trees, and other vegetation. There is a rapidly growing understanding that honey bees are beneficial as pollinators and are good neighbors, even in crowded urban areas. The green movement strongly supports having bee colonies in cities. The Obamas even installed a bee hive on the South Lawn of the White House, which has greatly improved the image of beekeeping. New York City recently took a giant step forward when the Department of Health overturned a longtime ban on beekeeping within the city limits. Hopefully, other cities throughout the United States will soon take similar action. Bees should be a natural part of our environment in urban areas.

# CHAPTER 2

# The World of Honey Bees

You may not believe this—the estimated number of insect species in the world ranges from 2 to 30 million. Approximately 20,000 are species of bees—and less than 10 species are honey bees. In this book, we focus on one species, the common European honey bee, known by its scientific name, *Apis mellifera*, meaning "honey-bearing bee." European honey bees are not native to America. Prior to the seventeenth century they were found exclusively in the Old World. European honey bees were imported by early settlers in eastern North America around 1622. They rapidly spread westward by natural migration and by early settlers. Honey bees were novel to the Native Americans, who named them "white man's flies." Fossils recently discovered in Nevada prove that America once had a native honey bee—approximately 14 million years ago. For unknown reasons, this species became extinct. Other kinds of bees in the United States—bumblebees, carpenter, leaf cutting, mason, and so on—are interesting and beneficial, but they don't store enough honey to harvest.

All species of bees share some common traits. The main one is that their food is nectar and pollen from flowers; they are totally herbivorous creatures. Their ancient ancestors were carnivorous, but that's another story.

## Honey Bee Society

Honey bees are very social and gregarious.
Thousands live together in highly organized
colonies. There are three castes: queen, worker, and drone. The queen is a fully
developed female with huge ovaries that produce eggs at an unbelievable rate.
Normally, there is only one queen in the hive. She is the source of pheromones—
powerful chemical messengers that have profound effects on the physiology and
behavior of worker bees. Romantic literature exaggerates and misrepresents her image
as that of a leader, as if she were an intelligent, thoughtful creature with humanlike
cognition. No, she doesn't wear a cute little crown. And she is not a leader in any
sense of the word. Nevertheless, her presence is sensed at all times by the other bees
in the colony. Anytime she is removed from the hive, even for a few minutes, bees start
fanning and seem to become nervous. Are bees addicted to the queen's "perfume"? Are
these behaviors similar to withdrawal symptoms? If she is permanently removed from

**A honey bee is a magnificent biological machine, once you get close enough to see what it's all about. A bee has about as many body parts as you have.**

the hive, worker bees respond by building queen cells to rear a replacement queen (see Chapter 5, Reproduction within Colonies).

Worker bees are also females, but their ovaries are tiny and normally nonfunctional. Workers are smaller than the queen, and there are numerous differences in body structure, including the pollen baskets on their hind legs. Their stingers, which will be discussed later, are different, too. Honey bees seen foraging on flowers are always worker bees. Except for the queen and a few drones, workers comprise the entire colony population.

Drones are males. They are sparse, or even absent, depending upon the season and the population of the colony. The easiest way to recognize drones is to look for their huge eyes, which converge at the tops of their heads. Drones never visit flowers because they do not forage outside the hive. Their tongues are too short to reach the nectar inside flowers. Their physical structures are functional primarily during the act of mating while flying. This will get your attention: when a drone mates—requiring less than five seconds—his extremely complex penis literally turns inside-out with explosive force, making an audible pop. Then the completely paralyzed drone instantly falls to the ground to die. (Most contemporary books include a little sex, so I hope you enjoyed this brief diversion.)

**A drone (center) is easy to spot. His big eyes help him find a virgin queen during a mating flight high in the sky.**

# Nose-to-Nose with a Honey Bee

Bees are so small—less than an inch long—that their exquisitely designed bodies are difficult to see and appreciate. If you viewed one under a magnifying glass or binocular microscope—magnified perhaps ten or twenty times—you would be astounded. Basically, there are three major parts to a bee's body: the head, thorax, and abdomen. Bees don't have backbones; the body is totally encased in an external skeleton, a hard shell connected by delicate membranes at critical "flex points," such as joints in the legs or segments in the abdomen. A dense forest of microscopic hairs covers the entire body. The hairs vary in shape and function, depending upon their location on the body. For example, tiny hairs on the wing surfaces probably have aerodynamic functions. Most hairs over the body are fuzzy, branched like miniature trees, and functional in causing pollination by spreading pollen from flower to flower as the bees are foraging. Again, pollination by honey bees is necessary to produce one-third of the food that we humans eat.

**A hobby beekeeper should know something about the most important body parts of a worker bee.**

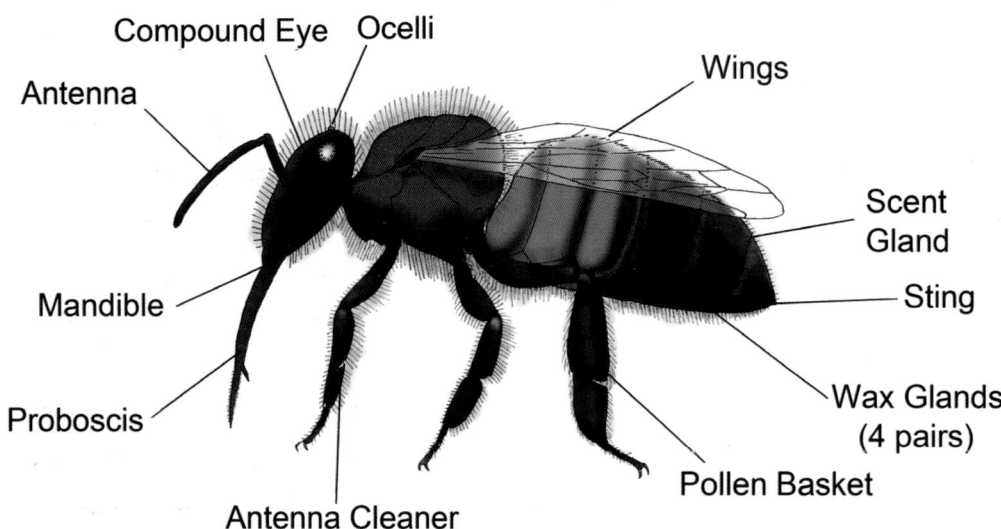

Compound Eye · Ocelli · Antenna · Mandible · Proboscis · Antenna Cleaner · Wings · Scent Gland · Sting · Wax Glands (4 pairs) · Pollen Basket

## The Head

Bees have two compound eyes, each composed of thousands of light-sensitive "micro-eyes" (*ommatidia*) that are fused together. Each ommatidium has a lens and a nerve connection. The ommatidia are connected to collectively generate a mosaic of sensory inputs into the bee's tiny brain, where the signals are integrated into a functional image. Yes, bees can see images—especially flower shapes—as well as colors. They see shorter wavelengths better than humans; ultraviolet is invisible to humans, but bees see

it as a color. Flowers are exquisitely endowed with nature's ultraviolet artwork, which we visually impaired humans can't enjoy. Foraging bees learn to associate images and colors of flowers with the reward of food (nectar and pollen).

*Left:* Not beautiful, but extremely functional, the bee's face shows antennae, an extended proboscis (tongue), and two of the three simple eyes on top of the head.

*Right:* Thousands of tiny eyes merge into a compound eye that can see colors and shapes. Notice the branched hairs, as well.

There are three additional simple eyes (*ocelli*) on top of a bee's head. Ocelli sense light intensity but not images. It's fun to speculate that ocelli also may be used for navigation during flight, considering their strategic location on top of the head and the possibility of a triangulation function. Why would a bee need three ocelli, instead of just one, to sense light intensity?

Two clublike *mandibles* aren't very impressive visually, but they get the job done when it comes to shaping beeswax into honeycomb. A long tongue (*proboscis*— pronounced pro-BAH-*sis*, not pro-BAH-*skus*) is used to suck nectar from flowers. It functions as a straw, yet it unfolds and retracts like a miniature landing gear. Extremely sensitive taste buds at the tip trigger the sucking response for the intake of nectar and water.

Two segmented feelers (antennae) on the head are multifunctional; they are used for tasting, smelling (even more sensitively than dogs), precisely sensing temperature to a fraction of a degree, and providing an astounding sense of touch. In the totally dark hive interior, the antennae function like a blind person's cane as the bees feel their way around their environment, contacting other bees as well as the combs. The antennae serve as calipers and are capable of measuring honeycomb specifications to the thousandth of an inch.

Here is an interesting concept to ponder: since the antennae are separated by significant space and each one is exposed to a different odor, do they "average" the two odors into one? Do they smell in stereo, just as we hear in stereo because our ears are separated?

*Left:* Try this at home: tasting with your toes—a routine feat for a worker's front feet (tarsi), which sense sweet liquids.

*Bottom Left:* Bees communicate using scents. They elevate the tip of the abdomen, expose the scent gland near the tip of the abdomen and a whitish membrane (moist with secretion), and disperse the scent by fanning.

*Bottom Right:* Worker bees secrete tiny, clear wax chips from four pairs of wax glands under the worker abdomen. Using only their sense of touch, they construct intricate comb inside the dark hive.

## The Thorax

This is the "engine room" where power is generated for locomotion—flying and walking. Powerful muscles fill the thorax interior. Bees walk on three pairs of tiny, multi-segmented legs that have some incredible features. *Tarsi* (feet) on the front pair of legs can taste liquids. When the front tarsus of a walking bee touches a sweet liquid, this stimulus triggers proboscis extension and a sucking response. From our perspective as humans, tasting with our feet requires quite a stretch of the imagination. A spine on each leg of the middle pair is used to dislodge pollen pellets that are transported on the hind legs. Two microscopic claws on each of the six legs provide traction when walking on textured surfaces. Foraging on flowers frequently requires walking upside down on waxy or smooth surfaces. This feat is easy for the bee. Whenever a surface isn't textured, the two claws on each tarsus fail to catch and simultaneously unfold a sticky membrane—located between the claws—that contacts the surface and sticks like a tiny suction cup. No one knows just how bees can release each sticky foot as they march along, upside down in flowers.

Bees have four wings, two on each side of the thorax. The two wings lock together along the wing margins by microscopic hooks, making them appear as a single wing. Navigation through the air looks easy, but it is terrifyingly complex, as demonstrated in slow motion studies of wing movements during flight. Around 230 wing beats per second create the typical buzzing sound associated with bees.

**Using blueprints transferred genetically, bees construct hexagonal honeycomb cells, which provide the greatest structural strength with the least building material. As a bonus, a bee fits into the cell, as this little one is demonstrating.**

### The Abdomen

Made up of external plates connected by flexible membranes, the abdomen can expand and contract like an accordion in response to the changes in volume of the internal organs (see the following section, Inside the Bee). Four pairs of wax glands on the underside secrete tiny chips of beeswax, which are used to build honeycomb. A scent gland near the abdomen tip releases orientation pheromones from a moist, white membrane that is exposed during use. The most exciting organ, by far, is located at the tip of the abdomen—scientifically termed the *sting*, but commonly called a *stinger*. (We'll call it a *stinger* in this book.) In Chapter 8, you'll learn how to overrule this little defensive weapon so you can truly enjoy beekeeping.

## Inside the Bee

The inner workings of the bee are almost beyond imagination. The bee's little body is crammed with functional organs similar to those of humans, just smaller in scale. A bee has a tiny brain—about the size of a pinhead—connected to a highly developed nervous system that extends the length of its body and into its appendages. On a cell-by-cell comparison, a bee's nervous system probably accomplishes far more than a human's does. Bees learn quickly and have fabulous memories; for instance, they can learn to associate the time of day with the onset of nectar secretion in their favorite flowers. They can also remember flower locations. Yet their behaviors are instinctive and reflexive, conveyed genetically from generation to generation. Contrary to popular opinion, a bee's complex behavior is not the product of intelligence and imagination. For example, bees do not learn, but inherit, the precise specifications for honeycomb construction.

**BEE BLOOD**

Bees have blood that is clear, not red like that of humans. Blood is circulated throughout a bee's body—even inside the tiny legs and hair-size antennae—by a complicated tubular heart.

Bees ingest only liquids or tiny particular matter, such as pollen grains, suspended in liquid. Their foods—nectar, honey, pollen, and water—are sucked up through the proboscis and pumped through a tiny tube (esophagus) to the abdomen. The tube connects to an expandable balloon-like structure—the honey stomach—used for temporary storage and transport of nectar and water to the hive. The honey stomach is incredibly distensible, capable of literally inflating the abdomen to nearly double its size and holding up to approximately two-thirds of the bee's total body weight. When artificial nectar (sugar syrup) is presented on a platter as a free meal, the bee can load up in about a minute. Nectar in the honey stomach—destined to be processed into honey—is later pumped out of the honey stomach, and deposited into honeycomb storage cells or transferred to hive mates.

**Sweat bees, which are attracted to salt in human sweat, come in a carnival of colors, including metallic green. They pollinate, too.**

Please note that *honey is made from nectar that never enters the true digestive parts of the alimentary canal.* Therefore, honey is not made from regurgitated nectar—an inappropriate description sometimes used thoughtlessly by beekeepers to explain the honey-making process, thus harming the image of honey. When a bee needs food for its own nutrition, a tiny one-way valve (*proventriculus*) at the back end of the honey stomach allows the controlled passage of tiny amounts of nectar into the true digestive stomach (*ventriculus*) of the alimentary canal.

Posterior to the ventriculus, and just before the anal opening, there is another expandable, balloonlike organ (*rectum*) that allows a bee to accumulate fecal waste for up to several months when it is confined inside the hive by cold weather.

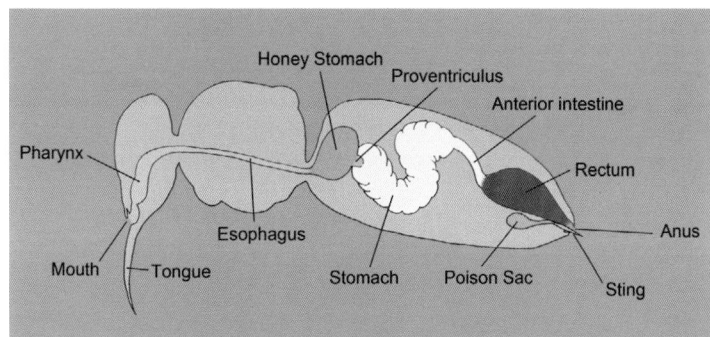

*Top:* The bee's alimentary canal contains a honey stomach (like a tiny balloon), shown here with a typical amount of reserve food that can be taken in small sips when needed through a one-way valve (proventriculus) into the digestive stomach for nutritional needs.

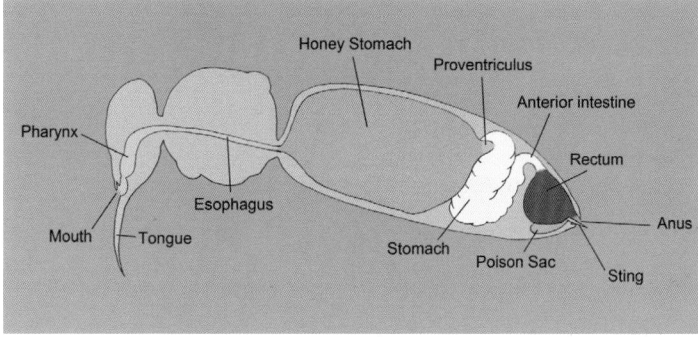

*Bottom:* The honey stomach expands while the bee sucks up a full load of nectar or water (weighing up to two-thirds of a bee's body weight) for transport back to the hive. A tiny, hairy filter in the proventriculus sifts out suspended pollen and transports it into the stomach to be digested.

# CHAPTER 3

# The Bees' Home

The bees' home is comparable in many respects to a human habitation. It provides protection from the elements, security, food storage, a place to rest, and a comfortable environment. It also provides a favorable setting for all members to communicate information about events inside and outside the hive that affect their survival.

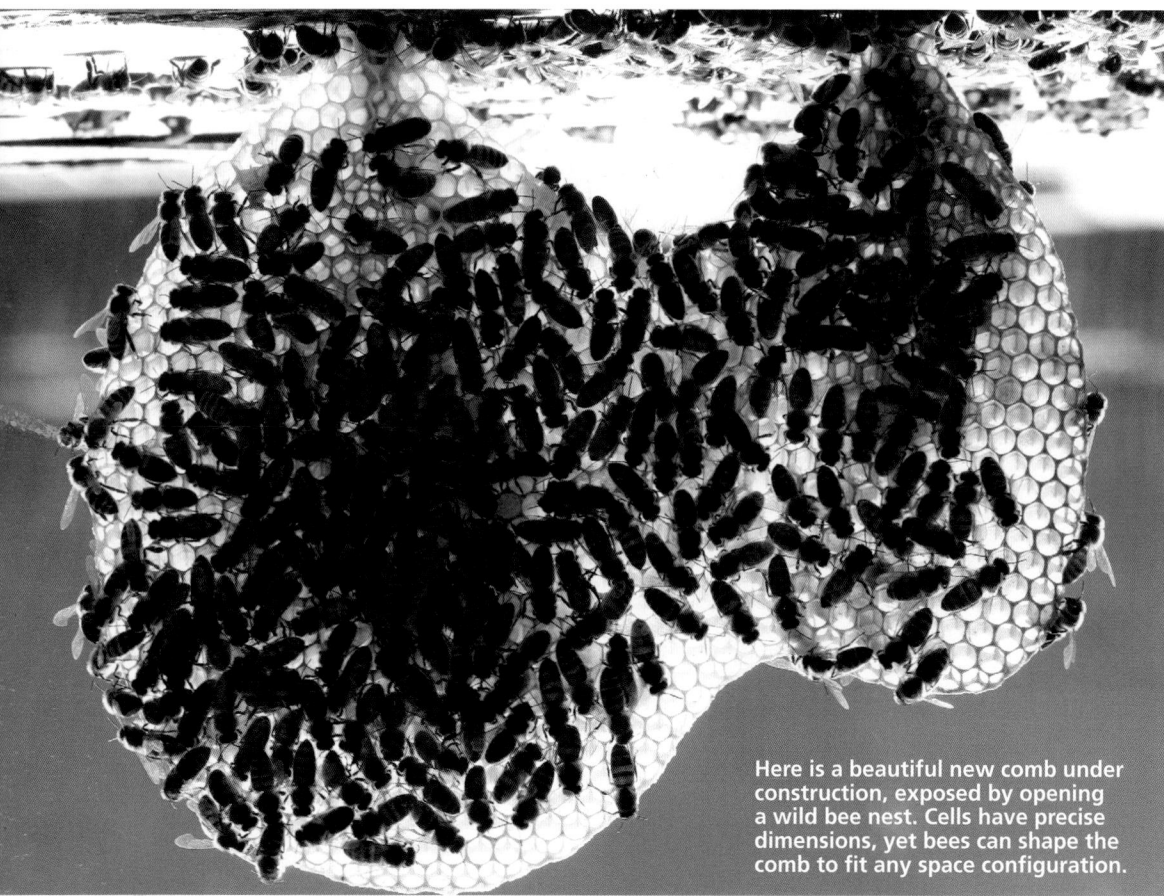

Here is a beautiful new comb under construction, exposed by opening a wild bee nest. Cells have precise dimensions, yet bees can shape the comb to fit any space configuration.

## Hive Structure

In nature, bees live opportunistically in a variety of natural cavities. When managed by beekeepers, they live in hives. The word *hive* refers to the manmade structure only. Once the hive is occupied by a living colony of bees, it is referred to as a *colony*. *Hive* and *colony* are used interchangeably, but most beekeepers refer to their hives as colonies. Beekeepers are constantly experimenting with different shapes and sizes of hives made of virtually any building material that is handy. Over the years, hive designs have changed dramatically.

The most important advance in manmade hives was a hive design in which bees are enticed to construct honeycomb in removable frames, thereby enabling the manipulation of combs, including the harvesting of honey, without destroying the combs in the process.

Bees aren't fussy about the shape of their nests. They easily adapt to cavities in the narrow and restricted spaces in building walls, hollow trees, chimneys, hollow support columns, and crevices in rock cliffs, to name a few. However, nest volume is important. They need enough volume for combs to accommodate brood rearing and storage of food (honey and pollen). Modern hives, at least in the United States, are composed of multiple chambers—boxlike in shape and typically made of wood. They do not resemble the dome-shape *skep*, or twisted-straw bee hive, still used in some parts of the world where wood is very expensive. Sometimes the skep is used symbolically to represent bee hives. For example, a skep is prominently displayed on the state seal of Utah, which is known as the "Beehive State."

## Nest Architecture

The architecture of the bees' nest is essentially the same whether in a natural cavity or a manmade container: multiple vertically oriented, side-by-side combs suspended primarily from the top. The approximate 5/16-inch space between combs permits the thousands of bees to move freely about or to cluster tightly, depending upon the temperature. L.L. Langstroth gets credit, at least in the United States, for discovering that this space must be incorporated between parts of manmade structures placed inside the hive. Otherwise, bees seal smaller spaces with *propolis*, a substance used by the bees as cement (see Propolis Foraging in Chapter 7), or fill larger spaces with unwanted combs. This makes it very difficult to remove

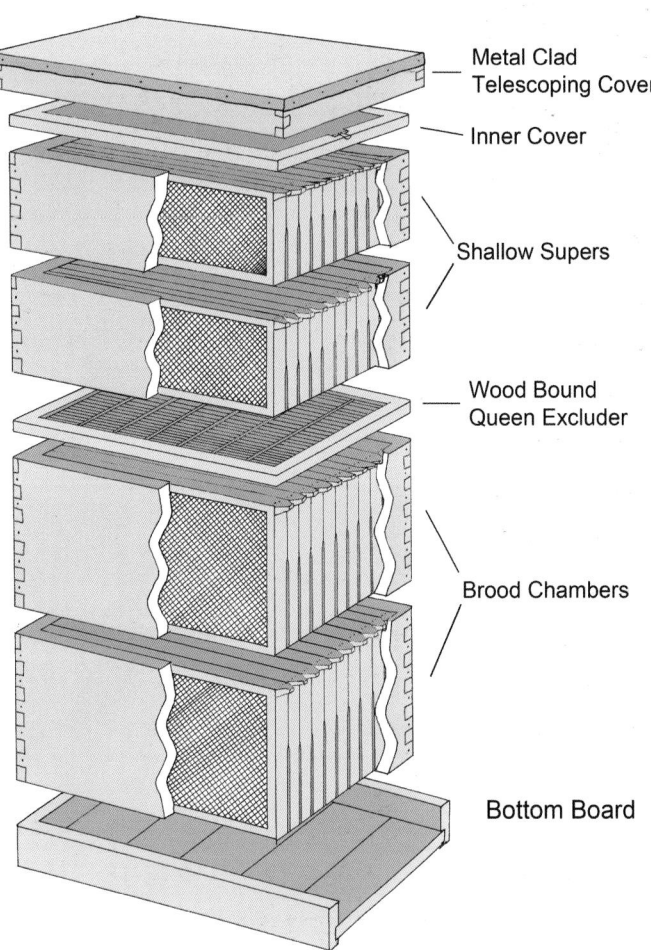

**Metal Clad Telescoping Cover**

**Inner Cover**

**Shallow Supers**

**Wood Bound Queen Excluder**

**Brood Chambers**

**Bottom Board**

**This is a cross-section of a typical hive, showing boxlike chambers of combs contained in rectangular wooden frames spaced apart as bees do in natural nests yet easily removable for inspection and honey harvesting. The number of chambers and supers varies according to seasonal and population demands. The queen excluder, with precisely spaced parallel wires, confines the larger queen (and thus all brood rearing) to the brood chamber but allows the passage of smaller workers to store honey in supers.**

combs to permit routine inspections of the nest or to harvest honey without seriously damaging the combs. Langstroth's discovery of *bee space* led to the design of modern hives and all the "furniture" used inside them.

Honeycomb (see Glossary) has one of the most amazing architectural designs found in nature. Bee behavior associated with comb construction is equally impressive. Bees don't need blueprints, as specifications for the comb are embedded genetically and expressed instinctively. The bees remove little flakes of wax from their wax glands and meticulously chew and shape them into hexagonal cells on both sides of the comb. The cell walls of natural honeycomb are several thousandths of an inch thick, are very smooth and uniform, and have precise dimensions. Amazingly, combs are constructed by sense of touch in total darkness. Man has made significant progress in the use of plastics to help bees make stronger and more durable combs (see Hive Equipment in Chapter 4).

**A skep is a dome-shape hive made of woven straw that allows bees to construct combs as in natural nests—irregular in shape and firmly attached to interior surfaces. Inspection and honey harvesting involves destruction of combs and there is little flexibility in expanding the nest volume seasonally. Primitive by modern standards, skeps are no longer used except in very poor countries where wood is prohibitively expensive.**

## BEE-INSPIRED DESIGN

Highly educated engineers with impressive degrees have confirmed that the hexagon provides the greatest structural strength for the least amount of building material. That is why hexagons are used in airplane wings, car radiators, shipping boxes, and myriad other manufactured goods. It's humbling to think that bees invented this shape millions of years ago, long before human applications.

# Components of the Hive

In modern hives, bees build combs in wooden frames designed to hang side by side with bee space between them in all directions. Combs serve a dual purpose: the cells provide efficient storage of food (honey and pollen) reserves, and at other times, the same cells may be occupied by developing brood (eggs, larvae, and pupae). A cell's use is influenced by its location inside the hive. Bees tend to rear brood in the lower part of the nest and store honey immediately above the brood. In a two-chamber hive, the upper chamber will contain mostly honey and is called a *honey super*, or just a *super*. Hives are expandable vertically to accommodate fluctuating seasonal space needs for brood-rearing and honey storage. At some point, the hive may be four or more chambers high. There may be two chambers of brood on the bottom and two or more supers of combs for honey storage on top.

A *bottom board* serves as the hive foundation. It's simply a floor with an elevated rim of the same dimensions as the chamber above. At the hive front, a narrow slot

**Screened bottom boards are used by beekeepers who wish to monitor Varroa mite infestations by counting the number of mites killed by various treatments. The mites accumulate on the slide-out tray beneath a screen that keeps the bees above it.**

provides entrance, exit, and ventilation functions. The need to monitor Varroa mite infestations requires a screened bottom board. A *hive cover* serves as a roof. There are two popular styles of hive covers. One is metal-clad on top and telescopes downward over a flat "inner cover." The other is simply a flat wooden cover with cross braces, known as a *migratory cover*.

Beekeepers need to control the separation of combs containing honey from brood combs to simplify honey harvesting. They prevent the queen from laying eggs in the honey supers by inserting a *queen filter*—known as a *queen excluder*—between the brood chamber(s) and honey super(s). There are several styles of queen excluders, but they all share the same design principle—passageways large enough to permit worker bees to pass through yet small enough to prevent passage by queens and drones

**Critically spaced wires prevent the passage of the queen into chambers where brood rearing is not desired, such as honey supers. This tool has many applications.**

because they are larger than workers. One potential problem with queen excluders is that worker-bee traffic can be restricted to some degree if there is not enough space immediately above and below the queen excluder to provide easy passage. The styles with parallel wires in wooden frames usually have more uniform spacing and respect bee-space requirements. Consequently, they accommodate the passage of workers better than some of the more economical queen excluders made of plastic or metal sheets perforated with round or elongated openings. *Queen excluders are useful for multiple applications and are an essential tool for the hobby beekeeper.*

Hives are usually painted white to reflect heat, especially in hot climates, but any color you like is just fine. Multicolored hives help bees find their hives when numbers of identical hives are close together, but if you just have a few hives with generous space between them, confusion concerning hive identity is not a problem. Apply high-quality exterior house paint to a primer coat. Some beekeepers like to decorate the front walls of their hives with exquisite designs. However, bees are not picky about hive decorations. They make the same amount of honey with or without artistic embellishment.

**Although beekeeping is a science, art has its place, too.**

# CHAPTER 4

# Getting Started

At this point, you should have already visited an apiary, made friends with one or more experienced and competent beekeepers, and learned a lot about bee biology and behavior. Now it's time to start your own hive(s). You may be lucky enough to find a local bee-supply store. If not, there are plenty of bee-supply companies on the Internet. Most bee-equipment suppliers will be happy to mail a free catalog when requested. Catalogs will be helpful in choosing your equipment. Consider carefully your choice of a bee-supply company because you'll likely continue buying from the same source for many years. A good choice may be a well-established supply company that offers a wide range of products and has a reputation for quality merchandise as well as a longstanding record of service to beekeepers. Some of the suppliers have served beekeepers for more than a hundred years and publish very informative trade journals (see Resources) that are invaluable sources for hobby beekeeping information.

# What You'll Need

Which hive equipment and other beekeeping supplies do you need? There is a bewildering array of beekeeping supplies and equipment to consider. You need to choose those items that are truly essential.

## Hive Equipment

Bees are not particular. They don't care about keeping up with the Joneses. In natural settings, they nest in almost any aboveground, dark cavity protected by a small entrance. But *you* need a modern hive that is designed to facilitate easy removal of combs. Honey stored in combs in conventional wooden frames can be harvested easily, and the empty combs can be returned to the hive for more honey storage. In the old days, before modern hives were developed, combs—and sometimes entire colonies— were destroyed during harvest operations, as it was impossible to remove the combs without cutting them into pieces.

Once again, you need to collaborate with your beekeeper friends to get advice about buying and assembling hive equipment. There are many confusing options.

**Ever take a hungry kid to a candy store? Beekeepers who live near this store are very lucky. Honey, pollen, candle-making supplies, bee toys and other goodies lure beekeepers like flowers lure bees.**

## DON'T DIY

Please curb your do-it-yourself instincts if you have the urge to construct your first hive from published plans. There may be small variations in the dimensions compared to factory-made equipment, which will perhaps cause problems later when you need to purchase accessories or replacement parts. You should buy factory-made equipment that conforms to standard dimensions. Unfortunately, equipment from different manufacturers varies in dimensions and is not necessarily interchangeable. Mixing equipment sizes can cause aggravating problems such as unwanted comb construction between chambers where there is too much space.

Several choices have to be made at the outset, and mistakes made at this point may be difficult and expensive to correct later.

Will you use eight-frame or ten-frame equipment? Most beekeepers prefer ten-frame hives even though they normally use only nine frames per chamber once comb construction is completed in the established colony. Ten frames fit initially, when the equipment is new. However, the accumulation of propolis and wax where frames make contact changes the spacing. Tight-fitting frames are difficult to remove and increase the risk of injuring bees. In addition, there is a tendency for bees to underutilize the outside combs because they are squeezed too close to the chamber walls. You can install frame spacers, but they prevent necessary sideways movement of frames that are being removed. Propolis and wax accumulations alone are sufficient to space the frames properly.

Will you use standard deep chambers exclusively for brood chambers and honey supers? The advantage of using only deep chambers is interchangeability, because all of your equipment will be identical in size. Or should you choose shallow supers for honey and deep chambers for brood? Shallow honey supers are much easier to lift, especially when full of honey. If you decide to use deep chambers for both brood and honey, there are two ways to avoid injury when lifting the chambers. One way is to work with a buddy to share the load; the other is to transfer combs individually to a temporary holding chamber, perhaps an empty deep chamber, during the hive inspection. Using this strategy, you can lift just one frame at a time—or as many frames as you can easily lift at one time when they are placed in the temporary holding chamber.

Perhaps the best hive configuration for the hobby beekeeper is two deep, ten-frame chambers for brood rearing and at least three shallow supers for honey storage. At the very least, you'll need a bottom board, one deep chamber containing ten sheets of comb foundation, and a hive top or cover when starting a new colony. It's smart to buy a second deep chamber at this time because it will be useful when feeding sugar syrup to your new, single-chamber colony (see Feeding Sugar Syrup in Chapter 9). Within one to two months, your new colony typically will need additional chambers to accommodate normal expansion, so you may want to buy all of the equipment for your hive at the beginning.

Another option is to purchase a beginner's kit, which should satisfy most of your needs when you first start beekeeping. Be sure that all of the components are really necessary by asking an experienced beekeeper for help, as the kit may include extra pieces that you don't need. In addition, some of the components may not be the best size. For example, the included bee smoker may be the small, economy-size type, which you'll have to replace later with a big one.

Honeycomb construction in modern hives is guided by sheets of comb "foundation" embossed with cell dimensions that precisely match those of combs in natural bee nests. Selecting the best kind of comb foundation for your application is a difficult decision. Pure beeswax foundation works well, but it's very difficult for beginners to install the sheets of foundation properly into the frames. Beeswax foundation must be reinforced with horizontal and/or vertical wires to provide strength and prevent the comb from warping or sagging. Also, the wires must be completely embedded into the wax. Some beekeeping-equipment suppliers sell the frames with beeswax comb foundation already installed, and this is the easiest solution.

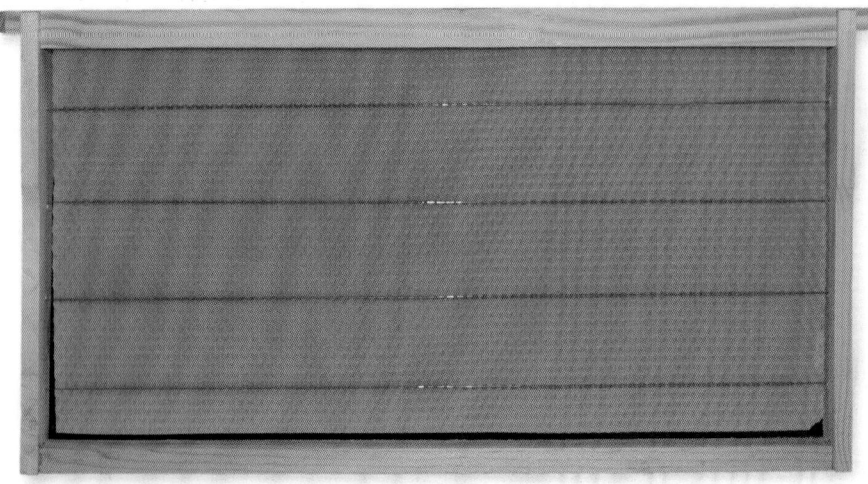

**This removable frame containing comb foundation with cell dimensions identical to natural honeycomb makes it simple for bees to build comb.**

Plastic comb foundation is becoming more popular and may be your best choice. Installation into the frames is much easier because wire reinforcement is not necessary. One of the best designs of plastic comb foundation has slightly

This plastic comb foundation with raised cell walls coated with beeswax helps the bees make a strong comb.

elevated cell side walls already constructed. Plastic foundation is not attractive to bees unless it is coated with a thin layer of beeswax.

Combs built on plastic foundation can be renewed after several years of use by scraping the comb cell walls away, down to the plastic foundation. This cleaning process removes disease-causing spores and microorganisms as well as chemical residue from various pest treatments. You don't have to remove 100 percent of the cells or wash the comb in any way, but you may need to recoat the plastic foundation using an old paintbrush dipped into hot beeswax, spreading it as thinly as possible. Beeswax is extremely flammable, so take precautions such as using a double boiler that is heated without an open flame, performing this operation outdoors, and having a fire extinguisher handy. The new beeswax coating will not be as neat as the original coating, but the bees will rebuild the comb nevertheless.

Occasionally, your colony will require supplemental food in the form of sugar syrup, especially when your new colony is just started and constructing combs. You

can choose from several styles of feeders to feed bees at the hive, where they can walk to the food source and collect a full payload in seconds. The most economical feeder is an inverted standard Mason jar (at least quart size), capped with a perforated lid. Bees suck the liquid through the perforations. The glass container reveals the liquid level so you'll know when to refill. (For feeding details, see Chapter 9.)

Hive-entrance feeders are convenient but frequently leak and may attract unwanted visitors such as ants.

A small hive, half the size of standard hives, is very handy at times. It's called a *nuc box* or *nucleus hive*. A nucleus hive is not a necessity when you first start beekeeping, but it's very useful for many applications. You can use it to start a new colony and then transfer the combs and bees to a standard hive later. It's also handy for maintaining a reserve queen throughout the active season to replace a failing queen in another colony. Portability and light weight combine to make it the preferred equipment when catching a swarm (see Catching a Swarm later in this chapter).

If your budget is stressed, you may want to delay the purchase of honey-harvesting equipment. There probably won't be enough honey to harvest for at least several months after you start your new hive. Bees need time to build all those new combs. During comb construction, an estimated 5–20 pounds of honey are consumed to produce 1 pound of beeswax.

The most essential beekeeping tool is a good bee smoker. A few puffs into the hive disrupts colony defense and puts you in control.

## Tools and Protective Clothing

You need more than just hive equipment. The indispensable tool you must have is a *bee smoker*. Buy a medium or large one for best results. Enjoying your hobby requires controlling the defensive behavior of bees during hive manipulations to minimize the chance of being stung. Beginners mistakenly assume that wearing protective clothing is their primary sting-prevention strategy—somewhat like a knight in armor. You *should* wear protective clothing for backup protection, but your first line of defense is smoke, which prevents most defensive behavior when properly used (see Chapter 8).

Other indispensable items are a bee brush and a hive tool. Aside from beekeeping applications, a hive tool is a great addition to your toolbox because it's so versatile and can be used for scraping paint, pulling nails, weeding in the garden, and more. Bees glue hive parts together using propolis. You'll need a hive tool for prying the hive chambers apart and removing frames one by one during routine hive inspections. Several styles of hive tools are available—buy the large size. You'll need all the leverage you can muster to separate the hive chambers.

Here's a bee brush in action. This handy tool is great for harvesting honey or removing bees from any surface where they are not wanted.

There are many styles of hive tools similar to this one. They all do the same thing—help you pry apart the hive chambers and combs.

**SEEING RED**

Don't buy tools or supplies in red if you can help it. If you must buy something in red, you may want to cover or repaint it. Red appears black to bees' eyes, and bees react defensively to dark colors. Why provoke them by waving a red hive tool at them?

Even though defensive behavior can be controlled to a great degree, you need protective clothing. Bee-supply catalogs display many styles. The main requirement for any kind of beekeeping is a veil that encloses your head. Coverall-style bee suits (also called jackets), especially those with built-in veils, are popular and very comfortable to use. If your budget is tight, you can use one of the veil styles that enclose just your head. You can then wear cotton trousers and a long-sleeved shirt if you remember to close all openings that expose your skin and to choose light colors.

Special bee gloves are available to protect your hands from getting stung, but you can use inexpensive cotton gloves in the beginning. After developing skills in the fine points of hive inspection protocol, most beekeepers enjoy the convenience and comfort of working without gloves even though they risk getting stung. Working without gloves should be your goal. There is no greater incentive to learn the delicate and precise manipulations of combs during hive inspections than to offer your bare hands as the test of your skills. This ensures that you will become a good observer of bees' defensive behavior so you can learn to avoid mistakes that may cause bees to sting. A compromise is to wear gloves with the fingers removed to provide freedom of finger movement yet provide protection for the other areas of your hands.

Using gloves presents several problems. Gloves become covered with sticky propolis and honey. They are uncomfortable in hot weather. When wearing gloves, you diminish your sense of touch and sometimes accidentally crush bees with your fingers when handling the frames. Beekeepers who always wear gloves rarely develop finesse in frame-handling skills, leading to rough handling of the combs and thereby compromising the safety of the bees, which can be accidentally crushed. Alarm pheromones released by crushed bees tend to stimulate alarm responses and defensive behavior by other bees. You don't want several dozen bees "angrily" flying around your head and body during hive inspections. However, despite the shortcomings of gloves, there are situations in which they are very useful—for instance, when bees are acting defensively—so it's a good idea to have a pair handy.

It's hard to beat a jacket with a built-in bee veil for comfort and protection. Head gear is a good idea because defensive bees react to exhaled breath.

# Preparing the Apiary Site

Consider putting your apiary site near the middle of your backyard, or at least as far away from your neighbors as is feasible. Ideally, hives should have direct sun exposure, but partial shade is acceptable. Preparation of the site before hive placement is helpful. You could place your hives on a cement slab. A more economical option is to spread asphalt roll roofing over the ground (use white to minimize heat). Be sure that the gravel-particle side faces up. This ground cover works well to control weeds, discourage certain insect pests (such as ants, small hive beetles, and termites) and protect the hive's bottom board from direct ground contact, which could cause rotting. You can make some sort of stand no more than 16 inches high—using bricks, blocks, or wood—to support the hives. Leave a minimum space of about 4 feet between hives.

Placing the hive stand on a ground cover, such as roll roofing, solves a number of problems. Notice the flat, wooden migratory hive cover.

# How to Get the Bees

There are several ways to acquire bees to stock your new hive. The key is to always start with at least 2 pounds of bees and a mated queen.

## Buying Bees for Your Hive

Once you have all the tools you need and the site is set up, you're ready to introduce bees into their new home. Some professional beekeepers specialize in producing bees to start new colonies. You can buy bees by the pound, along with a mated queen confined in a small cage inside the larger cage of worker bees. These are called *package bees*. You should stock your hive with 2 to 3 pounds of bees. There are approximately 3,500 to 4,000 bees per pound, depending upon how much food is in their honey stomachs. The best time to install package bees is early in the spring. It's a fun experience if you know the procedures, but it can be worrisome and stressful if you're unsure of yourself. You can find good video instructions on the Internet by searching for "installing package bees," but it may be a good idea to invite your beekeeper buddy to help you the first time you install bees.

## Catching a Swarm

An economical and entertaining way to stock your hive is to catch a swarm. Local beekeeping-supply stores and extermination (pest control) companies receive "swarm calls" from frantic homeowners who think that the end is near when thousands of flying bees suddenly appear on their property. The truth is that bees in a swarm are as gentle as they ever will be—not at all inclined to sting. Defensive behavior is associated with established colonies, not migratory bee swarms. Leave your name on file with these stores and companies early in the spring so that they can call you when they receive a swarm report. Before you commit to catching the swarm, you should try to verify (a) that it really is a swarm and not an established colony, (b) that it is large enough to justify your time and expense to capture it, (c) that the bees are newly clustered after just leaving their hive, and (d) that the bees are not in the process of moving into a new nest, where they will not be accessible for capture.

**Catch a swarm like this one and you can instantly populate your hive. It's easy to imagine that the bees are anxious to start comb building.**

Here are the key questions to ask. Did the swarm arrive within the last three days? If yes, proceed. Are the bees in a cluster at least as big as a football? If yes, proceed. Are they clustered less than 10 feet from the ground? If yes, proceed. Are the bees clustered on the side of a building or around a hole in the trunk of a large tree? If yes, stop here because the bees probably are already moving into a cavity where they won't be accessible. After the verification process, collect the swarm as soon as possible—before they migrate elsewhere and become a problem to someone.

There are many tricks to catching a swarm quickly and safely. This is the time to get help from your beekeeper buddy; you should not attempt to catch your first swarm without help. Also ask your buddy to bring an old comb or a used hive chamber to use as an attractant. If your hive equipment is new, especially if it is recently painted, it

probably will not be attractive—and may actually be repellent    to the swarm bees. Old combs and used hive equipment have highly attractive odors. You can return the borrowed

**Catching a swarm is a cinch when bees cluster low to the ground and you can shake them into used equipment with highly attractive odors. This beekeeper didn't really need protective clothing because swarming bees are docile, especially if you smoke them lightly. Beware: if they've been hanging around for several days, they gradually become defensive.**

equipment after the swarm has accepted the new hive and is starting to build new combs. Other tips and details are available on the Internet if you search for "how to catch a bee swarm." If you transport your new colony to your apiary location in an open vehicle, such as a trailer or truck, the blowing wind will cause the bees to stay inside the hive. If you decide to confine them, you must provide adequate ventilation. (See How to Move Hives in Chapter 9.)

## Getting Bees from a Friend

Another way to establish a new colony is to ask a beekeeper friend to donate 2 or 3 pounds of bees when they are seasonally abundant. Typically, this would occur during the spring, when colonies develop huge populations and are preparing to swarm by constructing queen cells (see Chapter 5). Take your new hive to the donor apiary and ask your friend to demonstrate the following procedures that are involved in transferring bees to your equipment. Find the queen in the donor colony, and put aside the comb containing her highness—outside the hive in a safe place—for the moment. She will be returned to the donor colony after the transfer of bees. If you want to produce a new queen from a queen cell, instead of introducing a mated queen, replace one of the frames in your hive with a borrowed brood comb from the donor hive that contains several capped queen cells that contain developing queens, one of which will produce a new queen for your colony. Queen cells are delicate, so be gentle.

Immediately transfer 2–3 pounds of bees—enough bees to densely cover at least five frames—from the "overpopulated" donor colony into your hive. Smoke the bees well, and then shake and brush them off the combs and into your hive (only violent shaking works). Once the bees have been transferred, close up the entrance with screen and temporarily replace your hive top with a ventilation screen that is secured to the top of the chamber during transit to your apiary (see How to Move Hives in Chapter 9). After securing the bees in the chamber, brush off any "hitchhiking" bees on the outside surfaces, spray the top screen with water to keep the bees hydrated, and transport them to your apiary. Immediately unload and place your hive in its previously prepared location, then smoke the bees, remove the screens, and replace the hive cover.

It is important to feed sugar syrup to your colony (see Chapter 9) during the first month or until all combs have been "drawn out" (the beekeeper's way to describe the completion of comb cell construction on the foundation). The sugar syrup gives them a good start, especially if nectar is scarce and the weather is unfavorable for foraging. Be sure to refill the feeder on

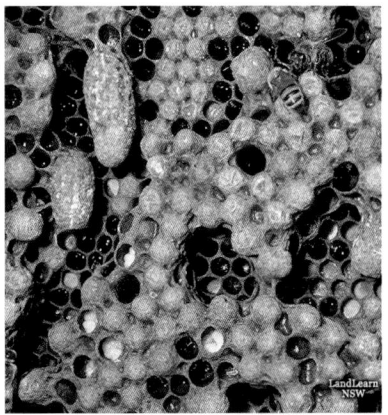

During the spring, strong colonies construct multiple queen cells—like the biggest cell shown here—just prior to swarming. Responsible beekeepers practice swarm prevention and control. Hint: make a quick, easy inspection for cells (without removing frames of comb). Tilt the upper brood chamber back to expose the frame bottoms where swarm cells point down like tiny fingers.

demand. You'll be amazed to see how fast and how much those little girls can eat. Clean the feeder between fills to prevent fermentation. Later, you can remove the bees from your friend's borrowed brood comb and return the comb to the original donor colony. Keep the comb warm during transport so the brood won't get chilled and die. When removing the bees from this comb (by shaking and/or brushing), be sure your new queen gets transferred safely into the hive. And don't forget to replace the removed comb with a frame of comb.

If you purchase a mated queen by mail before stocking your hive with bees, you could introduce her in a mailing cage, rather than introducing a frame of brood with queen cells, into your new colony. (See Queen Replacement and Introduction in Chapter 9.) A mated queen would start laying eggs almost immediately after she is released from the cage. The colony population would increase faster, which would increase honey production. *Introducing a mated queen instead of a queen cell is necessary, not just an option, if your hive is in an area where Africanized bees are established.* You wouldn't want your new home-produced virgin queen to mate with Africanized drones instead of European drones and produce a very defensive Africanized colony. Africanized bees are dangerous, especially in urban environments. (See Africanized Bees in Chapter 8.)

## Buying an Established Colony

You may be able buy an established colony from a beekeeper to jump-start your new hobby. One advantage is that you can immediately start honey production if there is an active honey flow (See Glossary). However, there are several disadvantages. An established colony is more challenging to manipulate because the sting risk usually is greater in colonies with larger populations. As a beginner, you could benefit from developing your hive manipulation skills, starting with a small colony that is not very defensive. You would also miss the thrill of observing the construction of new combs as well as learning about the growth of your colony from the very beginning. Furthermore, combs in the colony you buy may be old and may need replacement not too long after you buy them. Older combs may not be in the best condition as a result of various kinds of comb damage, including exposure to toxic chemical agents or medicinal treatments intended to control pests and diseases. An additional cause for concern is the possibility of accidentally importing serious bee diseases into your existing apiary. Despite the risks, most of which can be evaluated in advance by an experienced beekeeper's inspection, buying an established colony can be rewarding and economical compared to buying all new equipment.

## Hive Division to Start New Colonies

Dividing an established colony in half and then introducing a new queen into the queenless half is an easy way to create a new colony. *Divides*, as beekeepers call them, should be made in the spring. One option is to move the newly created colony to a distant location immediately after the division is completed. After a minimum of two weeks, you can return the new colony to the original apiary. If the introduced queen was not accepted in the new divide, don't try to introduce a replacement queen. She probably won't be accepted, because the physiology and behavior of workers changes during the time they are queenless. Simply combine the two colonies to make one.

Another option is to leave the newly created divide in the same apiary. Flight-experienced bees tend to return to the original, familiar hive location. You can attenuate the driftback problem by initially transferring about two-thirds of the bee population to the divide in the new location within the apiary. The populations will equalize somewhat as the flight-experienced bees return to the parent colony location.

## When to Expand Hive Size

As your colony grows, you must monitor its need for additional space. It doesn't take long for a new colony to outgrow its available space. When the bee population increases to capacity and all of the cells in the combs have been drawn out, you should add a second chamber. By this time, there should be enough bees of foraging age to sustain the nutritional needs of the colony, especially if there is a honey flow. If not, then you have the option to stimulate comb construction and population growth by feeding additional sugar syrup until a honey flow begins. Completing comb construction in the second chamber may require another month or so—maybe even longer. This time varies greatly according to weather, nectar availability, and colony population. During good honey flows, populous colonies can build combs quickly.

**This new comb is now ready to be filled with nectar or used for rearing brood.**

When the bees complete comb construction in the second brood chamber and you expect good foraging conditions for at least another month, then it's time to add your first honey super. Comb construction slows late in the summer and ceases altogether when cool fall weather arrives. If the expected honey flow doesn't materialize, the partially completed combs should be removed, emptied, and stored until the next season.

## EMPTYING HONEY FROM COMBS

Here's a simple procedure to empty partially filled combs while salvaging the honey. Place the super of partially filled combs on the bottom board. Spread the frames apart to provide at least double bee space between the combs. Scratch any capped cells to expose the honey. Now place a queen excluder on top of the super to prevent access by the queen. Place the two brood chambers, and any additional honey supers, above the queen excluder and replace the lid. Caution: be sure to cover the entrance with a robbing prevention screen because the honey fragrance is highly attractive to robber bees. During the next several days the nectar and honey will be transferred upward and stored in the chambers above because bees normally do not store honey below the brood nest. The success of this strategy is contingent upon available storage space, warm weather, and an adequate bee population. You can remove combs as they become empty and replace them with additional combs to be emptied. At the end of the procedure, when all combs are empty, you should reconfigure the hive by removing the honey super and queen excluder and placing the two brood chambers back on the bottom board. Now the empty combs are ready for storage.

# CHAPTER 5

# Honey Bee Reproduction

You will enjoy your beekeeping activities more and increase your management skills significantly if you understand the dynamics of reproduction in the colony. After all, how can you tell your children about the "birds and bees" if you don't know about the bees? One reason honey bees are so fascinating is that we love to solve the mysteries of their lives. It's humbling that we know so little about them, compared to what there is yet to learn. A good example is our longstanding sketchy knowledge about honey bee reproduction. Only about 50 years ago, I discovered queen mating pheromones that help drones find queens during the mating flight. Fortunately, we now know enough about honey bee reproduction to enable us to manage colonies efficiently.

# The Queen

All bees in a colony develop from eggs laid by the queen, so they all share the same mother. But, as a population, they typically have around fifteen fathers. Consequently, each bee has lots of half sisters. The queen mates with approximately fifteen drones and stores around 5 to 6 million sperm that will remain viable for her lifetime—averaging approximately one year. She releases several sperm as each egg is laid. Mating with several drones creates genetic variation within the colony, which is good insurance for a beneficial mix of genes. Virtually all aspects of a bee colony are under genetic influence: behavior, body coloration, disease resistance, life expectancy, foraging habits, and so on. Bees have survived and thrived for millions of years because this genetic diversity enabled them to adapt to changing environments.

Queen bees can actually be instrumentally inseminated to maintain their pedigree. A touch of carbon dioxide anesthetic, a quick injection of semen from the syringe, and the deed is done. A breeder queen bee may even be worth several hundred dollars.

The queen has no control over the drones that inseminate her. She mates while flying, never inside the hive. This open-air sexual extravaganza lasts only a few minutes. She mates with random drones from other colonies within a sizable area—at least a mile in radius—from her hive. Drones fly during the afternoons when the weather is warm and the wind is not too strong. They assemble in large numbers in specific locations, called drone congregation areas, where they hover and buzz around 20–50 feet above the ground, awaiting the arrival of the flying virgin queen. Large drone populations ensure quick mating and probably protect the queen from predation by birds and insects. Drones detect the queen instantly by smelling her special perfume—pheromones emanating from various glands in her body. They aggressively compete to assume the mating position on top of her abdomen. When she opens the tip of her abdomen, the drone's penis instantly inflates with explosive force into her vaginal chamber, depositing his semen in less than a second. Now paralyzed and dying, the drone falls to the ground and becomes a feast for other hungry critters, especially ants.

Other drones follow in quick succession and enjoy the same fate. When she gets enough sperm, she heads back to the hive. Mating flights average around thirteen minutes. Some queens take one or more mating flights on subsequent afternoons. The approximately 5 to 6 million sperm migrate into a tiny sperm storage organ (spermatheca), a sphere not much larger than the head of a pin, where they remain for the queen's life.

Positions assumed by drone and queen during mating:

Initial mounting position.

Intermediate position.

Drone separation by genital explosion.

# Reproduction within Colonies

Queens have the option of fertilizing each egg as it is being laid. Almost all eggs are fertilized and are laid in worker-bee cells. During the active season, queens typically lay around 1,200 eggs daily. Worker bees are female and develop from egg to adult in approximately twenty-one days, at which time they chew their way out of the captive cells

These worker eggs, enveloped by an ethereal view of the hexagonal cell openings, are stuck on end when they are laid and gradually lean over until they hatch three days later, revealing larvae almost too small to see without magnification.

to join the community. When colonies are preparing to swarm, the queen will lay fertilized eggs in approximately six to twelve queen cells, which are oriented vertically. Queens develop from egg to adult in about sixteen days. Drones are male and are produced when the queen lays unfertilized eggs in drone cells, which are larger than worker cells. Full development from egg to adult occurs in twenty-four days. How can unfertilized eggs hatch? The process is called parthenogenesis, a phenomenon that occurs in other animals, too, including some species of fish, insects, frogs, and lizards.

Drones are produced in large cells with protruding cappings. Worker cells are smaller and their cappings are smooth.

Young larvae in the open cells are floating in a C-shape on white brood-food secretions from nurse bees.

As each tiny sausage-shape egg is laid, it sticks at one end to the cell bottom. Eggs for all castes hatch three days after they are laid, with each egg yielding a tiny white larva that is almost invisible without magnification. Nurse worker bees deposit nutritious secretions resembling dilute mayonnaise from their brood-food glands into the brood cells to nourish the worker larvae during the first three days of larval life. Then nurse bees modify the worker diet to include pollen. Queen larvae are fed royal jelly throughout larval development, providing a nutritional stimulus that causes them to develop into fully functional females with large ovaries. Although workers are also females, their ovaries are tiny and typically nonfunctional.

Larvae grow at an astounding rate in the warm brood nest, incubated at 93–94 degrees Fahrenheit. Owing to the constant temperature, the various developmental stages—egg, larva, and pupa—develop at predictable rates for each caste (see table Honey Bee Development). It's very important to know, and to be able to recognize, the life stages of developing brood so you can play detective in determining time-related events in the colony.

## HONEY BEE DEVELOPMENT
(Days After Egg Was Laid)

| Stage | Worker | Queen | Drone |
|---|---|---|---|
| Hatching | 3 | 3 | 3 |
| Cell Capped | 8 | 8 | 10 |
| Becomes a Pupa | 11 | 10 | 14 |
| Becomes an Adult | 20 | 15 | 23 |
| Emerges from Cell | 21 | 16 | 24 |

Colony population is of paramount importance. Sparsely populated colonies simply struggle to survive, collecting just enough food to sustain their activities. They are a liability to the beekeeper. Highly populous colonies are efficient pollinators and produce lots of honey quickly when nectar is available. The number of eggs laid daily and the average life expectancy of the bees determine the population in each colony. Workers live for around five to six weeks during the active brood-rearing season. If the queen lays 1,200 eggs daily and the average worker bee lives thirty-five days, the population would be 42,000 under ideal conditions. Mortality from various causes—including diseases of the developing brood, predation while foraging, and contact with toxic substances, such as pesticides in the environment—would reduce the population.

## REPLACING AN OLD OR UNPRODUCTIVE QUEEN

To replace an old queen with a new one, you must first remove the old one so that she can be sacrificed. Sometimes you won't be able to find the queen by conventional searching, inspecting each comb until you finally see her. An alternative search strategy is to filter her from a large worker population using a queen-filter chamber, made as follows. Attach a queen excluder to the bottom of an empty deep chamber. Line the interior chamber walls with sheet metal or heavy-duty aluminum foil. Lubricate the lower 5 inches using a thin layer of petrolatum jelly or vegetable oil. Remove the upper brood chamber. Place the filter chamber on top of the lower brood chamber. Shake and brush all of the bees—and hopefully the queen, too—into the filter chamber, and smoke the bees downward into the lower brood chamber. In a short time, the queen will be exposed on top of the queen excluder—easy to see. If the queen doesn't appear, then she must already be in the lower brood chamber. If so, place the chamber of empty brood combs on the bottom board and repeat the filtration process with the second brood chamber. In the process, be sure to glance at the bottom board for a "runaway" queen. If you insert a second queen excluder between the two brood chambers four days before the operation, you'll only have to filter one chamber—the one containing eggs. Once the old queen is removed from the colony, you can replace her with a new queen.

Healthy, young, vigorous queens with the appropriate genetic credentials ultimately determine colony populations by virtue of the numbers of eggs they lay, as well as the longevity of workers as influenced by their genetics. Your success as a beekeeper depends on your queen-management skills. Learn how to judge queen performance. She must be producing lots of eggs—consistent with seasonal constraints—that yield a compact pattern of brood with few open cells in the capped brood areas. Furthermore, the colony should not be unusually defensive when manipulated properly. Do not hesitate to replace marginal queens at any time during the active brood-rearing season.

Seasonal changes affect brood-rearing activities. Up to 2,000 eggs may be laid daily in the spring, when colonies are preparing to swarm. During the summer months, the rate of brood-rearing stabilizes. As fall approaches, egg-laying diminishes until late fall or early winter, when it stops completely in colder climates or continues at a low rate in southern states. Brood-rearing resumes again in early January, even when the hive is covered with snow, and low temperatures (below 55 degrees Fahrenheit) prevent flight. How can a coldblooded insect maintain normal brood-nest temperatures of 93–94 degrees Fahrenheit during very cold weather? They eat lots of honey, flex their muscles to produce heat metabolically, and conserve heat by clustering tightly together. Their bodies provide insulation and literally function as a warm blanket over the brood. Bees on the outside of the cluster exchange places with bees inside the cluster to stay warm. Some bees, known as heater bees, specialize in producing heat. During the winter and early spring, when nectar foraging is not possible, it's not uncommon for a hive of bees to consume 60 pounds of honey. Ironically, the bees that made honey for winter stores died long before winter arrived.

## Production of New Colonies

Colonies may die owing to events such as starvation, diseases, parasites, and insufficient populations to survive the winter. In their natural setting, lost colonies are replaced by new colonies produced by swarming, an event that almost always happens during spring for European honey bees. The parent colony divides into two populations, and about half of the bees leave to establish a new nest, accompanied by the old queen. The "stay-at-home" bees produce a new queen to continue reproduction.

The act of swarming is perhaps the most dramatic event in the lives of honey bees. Here's how it happens: Egg production increases dramatically in response to warming spring weather as well as an abundance of pollen and nectar from spring flowers. Within a few weeks, the colony population essentially doubles. Multiple queen cells—usually at least six—are constructed in the brood nest. A few days prior to the emergence of a virgin queen, the old queen's ovaries begin to shrink. Egg-laying essentially stops, and she loses enough weight to permit flight for the first time since her mating flight.

Soon there is not enough space in the hive to accommodate the population. On a warm, sunny spring day, the bees that will swarm engorge on honey. Without warning, there is a sudden, almost explosive exodus of hyperexcited bees from the hive entrance. A gigantic cloud of buzzing bees fills the air—an awesome spectacle for the uninitiated. They migrate a short distance and collect in a cluster nearby, usually on a tree branch. On rare occasions, they cluster in a public place, such as on the scoreboard at a major sporting event, providing the media with an opportunity to sensationalize the imaginary risks of bee stings. Scout bees from the swarm search the area and communicate the locations of potential nest sites by performing wagtail dances (see Bee Dances in Chapter 6) on the swarm cluster surface. The best sites stimulate the most active dances. When most of the scouts indicate a common location—an agreement of sorts—a wave of excitement sweeps through the cluster, leading to another massive flight, this time to the new nest site. Upon arrival, they immediately enter the scouted cavity, fanning and releasing orientation pheromones as they enter. They start building comb within hours. The queen is lavishly fed; within a day or so, she will resume laying eggs.

## PASSING THE CROWN

Discovering fewer than six queen cells in an active brood chamber usually indicates that queen supersedure, rather than swarm preparation, is underway. Sometimes the old queen and the superseding queen, now laying eggs, will coexist for a few weeks before the old queen dies.

Within approximately three to four months, the new colony will have become established and will contain a full complement of combs, stored honey and pollen, and a large population, all of which is required to survive the winter. In time, the old queen will be replaced, superseded as she succumbs to old age.

**A new queen emerges from her cell. Successful emergence is indicated by a smooth perimeter at the cell opening, cut by the queen to make an "escape hatch."**

What happens in the parent colony after the swarm departs? Multiple virgin queens usually emerge from their cells around the same time. Rival queens engage in fierce stinging attacks until only one virgin queen remains. Virgin queens also initiate the destruction of capped queen cells containing their younger counterparts and sting them before they can complete development. This is the only time queens ever use their stingers. Queens never participate in colony defense, and they will never sting you when you handle them.

The victor virgin queen will be sexually mature and ready for her mating flight about a week after she emerges from her cell. Weather permitting, she may mate by the time she is ten days old. The time window for mating flights—according to the queen's biological clock—is approximately seven to twenty-one days after her emergence from the cell. If she is unable to mate within this time frame, perhaps because of extended bad weather, she will eventually lay eggs, but they will be infertile—producing drones exclusively—and laid in worker cells. As they develop, these cells produce cappings that protrude outward noticeably farther than capped cells that contain developing worker bees. A colony with an infertile queen is doomed because there will be no young bees to replace the population that is dying of old age.

## How Long do Bees Live?

Queens live for approximately one year—much longer than workers and drones, which live for about five to six weeks. A few queens live for as long as two or three years, but old queens are a liability to the colony due to diminished egg-laying capacity, a principle cause of reduced colony populations and reduced honey production. Their performance usually diminishes long before they die, similar to humans.

Egg-laying capability is not the only measure of a queen's performance. Queens produce pheromones that greatly affect the activities, especially foraging activity, of workers. Pheromone production diminishes in quality and quantity as queens age. Poorly performing queens should be replaced any time during the active season. To anticipate and avoid problems associated with declining functionality, queens should be replaced at least once a year. If honey production and efficient pollination are the reasons you keep bees, your hives should not become assisted-living homes for old queens. Most hobby beekeepers have tender feelings toward their queens and typically are reluctant to sacrifice poorly functioning queens. Consequently, they sacrifice colony welfare and productivity.

One way to monitor the age of a queen is to paint an identification mark or glue a plastic identification tag onto her thorax when she first starts laying eggs. Use a quick-drying paint such as model airplane dope or fingernail polish. White is probably the easiest color to see. Yellow resembles pollen loads, so it is not as distinctive. Practice on drones or workers before you risk your valuable queen. You can safely handle drones because they don't have a stinger. Dip a very fine paintbrush—or maybe the tip of a toothpick or the end of a straightened paper clip—no more than an eighth of an inch deep into the paint and gently apply it to the top of the thorax. Take great care not to apply too much paint. The paint must be thick enough to penetrate the hairs and not just stick on top; hairs become the anchor for the paint. You can gently wiggle the applicator a little to get better penetration. The paint must be applied only in the center of the thorax top—not at the wing bases or on the neck. Allow the paint to dry a moment or so before placing the queen back into her cage or releasing her into the colony.

Drones are typically found in colonies during the spring and summer months. Their only function is reproduction; they do not forage for food. They become sexually mature and start mating flights when they are about ten days old. During the fall, they are physically "evicted" by the workers—cast outside the hive to starve to death. Mother Nature decided a long time ago that drones would jeopardize the colony's chances of survival by eating valuable winter honey reserves needed by the workers.

Worker bees live for approximately five to six weeks in the spring and summer. Those reared in the fall live for several months—long enough for the colony to survive the winter—and are replaced by young bees in late winter or early spring.

## ALTERNATE ID

Some beekeepers prefer to monitor the queen's age by clipping her wings on alternate sides for alternate years. Wing clipping will not prevent swarming. In this situation, bees will swarm with a virgin queen rather than with the old, clipped queen.

Surviving the winter is a challenge, but bees cluster to stay warm.

# CHAPTER 6

# Activity Inside the Hive

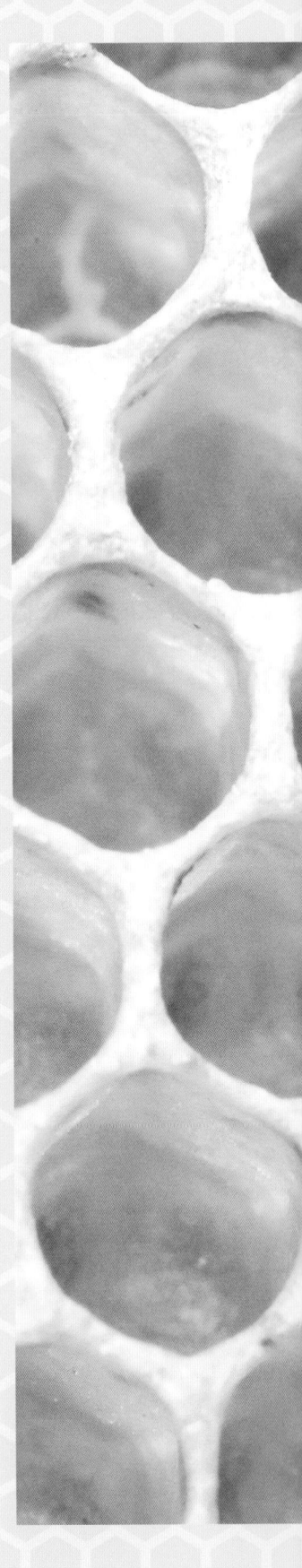

**W**e take for granted our wonderful ability to imagine things. This is a great gift. Conversely, one of humanity's greatest faults is our frequent failure to discriminate between the scientific reality and what we've imagined. Beekeepers tend to regard bees as if their little brains are similar in function to human brains. This assumption can be a handicap to beekeepers who are unable to accept bees as relatively primitive animals. The activities and behaviors of all insect species are similar in that they are controlled genetically, the insects having inherited biological "software" for their computerlike brains during millions of years of evolution.

## Do Bees Think?

Bee activities are not directed by a thinking process, or cognition, similar to that of humans. But how can they perform complex behavior, such as constructing honeycomb to such precise specifications, if thinking is not involved? Under experimental conditions, if capped brood is removed from a hive and confined in an incubator, the adult bees emerge normally. While in confinement, they have no opportunity to learn from older bees. When these naïve bees are used to establish a new colony without older bees, they soon spontaneously perform all of the activities that their sisters are performing in the parent colony. Their behavior is dictated genetically, hardwired into their DNA. They simply respond to the stimuli in their environment. They are robotic. Humans understand the concept of survival, but bees don't even know what survival means. They have no emotions, no imagination, and no foresight. Bees don't hold grudges, they don't behave any differently to strangers compared to their beekeeper owners, and they don't make honey because they like you. Anthropomorphic tendencies are counterproductive to rational beekeeping strategies. However, your bees are for your entertainment. Believe what you wish. If you find it enjoyable, you can even talk to your bees. But get some professional help when they start answering!

## Observing Bee Activities

You've heard the expression "busy as a bee" many times. It perfectly describes the honey bee colony. That's a major reason why honey bees are fun pets. Most animals have long periods of inactivity. Observing a snake hibernating in a terrarium is the antithesis of watching bee activities inside a glass-walled hive. Yes, you can mount a glass-walled

**An observation hive is probably the most underutilized beekeeper's toy. Peeking inside the bees' world through a glass wall let's you see myriad activities, day and night, without disturbing them and provides endless entertainment.**

observation hive inside your home. Such hives can also be mounted in other public places. An observation hive can be connected by a tunnel from its entrance to the outside world so it can forage normally. You can observe myriad bee activities any time of the day or night. You can see normal behavior, as opposed to the abnormal behavior you usually see during routine hive inspections. When you open a hive, the smoke, vibrations, harsh sunlight, disruption of the delicately controlled environment, and exposure to human odors disturb the bees, producing "disturbed" behavior.

First-time observers of bees inside the colony are awed by the crowded conditions: thousands of bees densely packed in such small spaces, walking over each other and endlessly interacting physically and chemically. Without doubt, claustrophobia is not an issue for bees. Many fascinating behaviors are happening simultaneously. Colony defense (stinging behavior) is probably the most important behavior for the beginning beekeeper to study. Your skill, and especially your finesse in controlling bees' defensive behavior, will determine your success as a hobby beekeeper. Preventing stings is so important that Chapter 8 is devoted entirely to the subject.

## Temperature Control

Honey bees are cold-blooded animals, yet they can maintain the brood nest temperature inside the hive at a constant 93–94 degrees Fahrenheit, even when ambient temperatures outside the hive range from -30 to 120 degrees. Tiny temperature-sensing cells found in the terminal six segments of the bee's antennae monitor the interior hive temperature. Whenever the ambient temperature inside the hive gets too warm, some bees forage for water and spread tiny droplets throughout the brood nest area. Fanning bees circulate air in directionally controlled air streams throughout the brood chamber. Bees "invented" *evaporative cooling*, an efficient and economical procedure sometimes used in human habitations, especially in hot, arid climates.

Bees don't heat the entire volume of air inside the hive. The critically incubated area where brood is being reared is somewhat spherical in shape and concentrated in the center of the brood chamber. Clustered nurse bees generate metabolic heat, which is complemented by additional heat produced by developing brood. Just outside the incubated brood sphere, the temperature fluctuates greatly according to the ambient temperature of fresh air entering the hive.

## Grooming Behavior

Worker bees spend a lot of time grooming their bodies. They have a fringe of tiny, stiff hairs—resembling the teeth of a comb—on the inside margins of their legs that can reach and comb the hairs on virtually all body surfaces. Cleaning their antennae is especially critical to keep the thousands of microscopic sensory cells free of debris. On each front leg, there is a small semicircular notch that perfectly matches the diameter of their antennae and is lined with *comb hairs*. Bees literally comb the antennae by reaching up with their leg and pulling the antenna through this notch, called the *antenna cleaner*. Worker bees also groom each other. A "dirty" worker does a little dance to solicit grooming by nest mates. She stands stationary and vibrates her body from side to side about four to nine times a second. Mutual grooming is especially directed to the base of the wings, a location that the dirty bee can't reach. The bees' behavior is reminiscent of monkeys picking parasites from each other—you scratch my back, and I'll scratch yours!

**Residual brown cocoons outline the egg-shape, incubated brood-rearing area in the center of this frame that was used in the brood chamber. Each comb removed from the hive represents a cross section of the brood nest.**

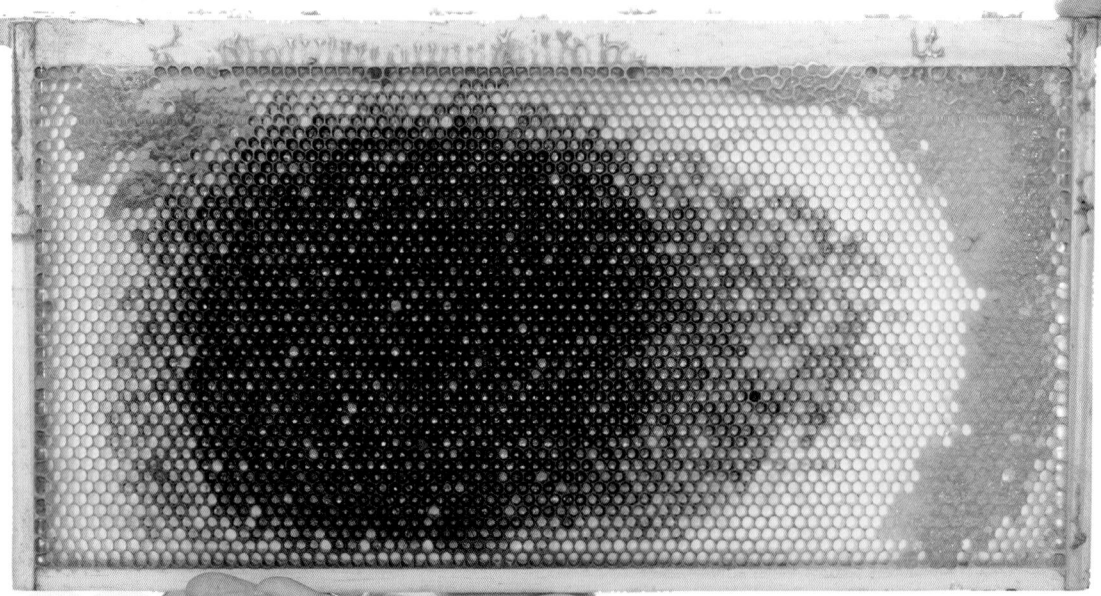

# Queen Pheromones and Behavior

The name "queen" indicates that this reproductive female is the leader and implies that she has special "duties" and powers over the colony population. Actually, she does unknowingly influence the physiology and behavior of workers by virtue of chemical messengers—pheromones—that are secreted from various parts of her body. The most obvious responses of worker bees can be observed in the immediate vicinity of the queen. When the queen moves near the workers, the workers' behavior immediately changes. As she moves along, workers move out of her path as if repelled. If she stops, they encircle her and start to groom and feed her. This is known as *retinue* behavior. You can elicit nearly identical behavior from workers by exposing them to pheromones that have been removed from a dead queen and applied to a "dummy" queen that bears no physical resemblance to the living queen. This proves that the workers are responding to pheromones, not to an individual regarded as a leader. We can monitor these behaviors in a glass-walled observation hive or on an exposed comb, but research to date indicates that workers inside the totally dark hive sense the queen's presence by her odor.

The dynamic and complex interactions between a queen and workers can be understood best by visualizing spatial zones near the queen where multiple pheromones produced on different parts of her body are constantly changing in concentration according to the proximity of workers to her body. The sensitivity of workers to the pheromones will also vary according to the age and physiological condition of each worker. These complex, transient interactions defy accurate definition but can be presented graphically.

**These multicolored zones represent dynamic changes in concentrations of pheromones, which stimulate different behaviors in the worker bees surrounding the queen.**

If a queen dies or is removed from the colony by the beekeeper, the absence of her pheromones causes workers to produce "emergency" queen cells from existing worker brood. A few cells are lavishly fed with royal jelly to the extent that the larvae actually "float" into the enlarged worker cell, now pointing downward. The resulting virgin queen mates and replaces the original queen. The interruption in brood-rearing lasts approximately three to four weeks.

If the virgin queen is lost, as sometimes happens during the mating flight, then the colony cannot produce a new queen and becomes "hopelessly queenless." Within a month or so, in the absence of queen pheromones, ovaries in some workers become enlarged and capable of producing a few unfertilized eggs. These bees are known as *laying workers*. The unfertilized eggs develop into drones. A laying worker colony is easily identified by (a) a greatly reduced colony population, (b) multiple eggs per cell, some of which are in random positions, even attached to the side walls of the cells, (c) the absence of normal worker brood, and (d) scattered cells that contain drone brood in worker cells with protruding cappings typical of drone cells. As the old workers die and are not replaced by new ones, the colony eventually dies.

## Nest Cleaning

After each adult bee emerges, worker bees immediately scrape and lick the interior brood-cell walls until they are very shiny and smooth. Queens are attracted to freshly cleaned cells—perhaps by worker pheromones—where they lay eggs to start the cycle over again.

Bees that die inside the hive are grasped by housecleaner bees, dragged out through the entrance, and dropped to the ground. Within several hours, depending upon weather conditions, their bodies lose enough weight by drying to be airlifted by housecleaner bees up to several hundred yards or more from the hive and then unceremoniously dropped like tiny bombs, creating a feast for critters on the ground. Studies show that, during the active foraging season, at least 90 percent of bee mortality occurs outside the hive.

Cleaning behavior removes disease agents from the hive environment and minimizes the attraction of scavengers and pests—especially ants—to the hive location. Bees will remove any foreign material, such as leaves or grass, that might be accidentally introduced into the nest during hive inspections.

## Comb Building

When workers are about twelve days old, their wax glands begin to secrete tiny flakes of beeswax. They chew the wax and fashion it into the architecturally complex honeycomb that functions as a vertical "floor" where almost all activities inside the hive take place. Newly constructed combs are light yellow in color and are one of nature's most artistic creations. As combs age, they become dark brown, owing to the accumulation of pigments from pollen as well as residual cocoons—the silky coverings with which larvae are enclosed during their transition to an adult state—

that gradually form a multilayered lining of the interior walls of brood cells, especially near the cell base. The cocoon layers trap disease-causing spores and organisms that may be released later—even years later—as bees are cleaning the cell interiors. Various impurities and chemicals gradually dissolve into the wax, rendering old combs unsuitable for brood-rearing. You should replace combs made with beeswax foundation when they have been used for approximately five years. (Be sure to record the year on top of the comb frame when it is new.) Beeswax is a very expensive product, energetically speaking. Bees have to eat an estimated 5–20 pounds of honey to secrete 1 pound of beeswax.

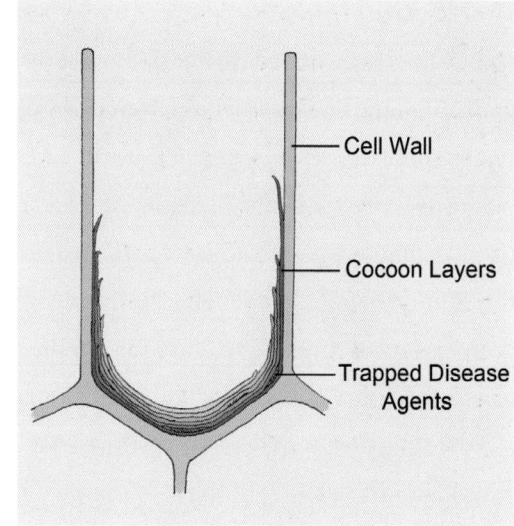

Cell cleaning by housecleaner bees is not perfect. Old, tough, residual cocoon layers can cover spores and other disease agents left behind in cells where mortality occurred. Subsequent housecleaning can expose disease agents that can start growth anew.

## Communication and Food Sharing

Efficient communication is the fabric of social behavior. It enables the thousands of bees in a colony to function almost as one organism—sometimes referred to as a *super-* or *supra-organism*, in which individual bees are compared to the individual cells of an organism. The status of each bee changes from moment to moment, and the summation of these changes is reflected in the overall colony condition. Vision plays no part in the communications inside the dark hive. The bees communicate by sounds, odors, taste, and touch. For example, when bees encounter each other, they instantly perceive the others' respective status. A hungry bee says, in her own special way, "Can you spare some food?" If the behavioral answer is yes, the donor bee spreads her mandibles and discharges a droplet of honey or nectar from her honey stomach onto her mouthparts. The hungry bee senses the food, extends her strawlike proboscis, and sucks up the food. Food sharing inside the colony is dynamic and continuous. Experimentally, if one bee were fed a drop of honey that contained a radioactive tracer, virtually every bee in the colony would contain the tracer within approximately forty-eight hours.

# Bee Dances

Honey bees have one of the most interesting communication systems in the insect world. Foraging bees returning from rewarding nectar and pollen sources perform recruitment dances on the brood-comb surface that are closely monitored by naïve potential foragers. These dances apparently communicate the distance and direction to the foraging location. Foragers also share nectar with the recruits, thereby providing flavor information. In addition, the fragrance of the flower source clinging to the dancers provides supplementary information. Within minutes, the newly recruited bees leave the hive independently and find the foraging areas indicated by the dances. After several round trips, they will perform recruiting dances, too, if the foraging rewards are sufficiently stimulating. If the food source diminishes in productivity—owing to excessive competition, shrinking nectar and/or pollen availability, adverse weather, and so on—foragers will stop recruitment behavior even though they may continue to collect food. Further reduction in food reward will temporarily terminate foraging activity. The net result is that the pace of foraging activities and resulting communication mechanisms are constantly adjusting to the dynamics of nectar and pollen availability. If the colony contains a normal foraging bee population, up to several hundred foragers can return per minute when nectar and pollen are plentiful. The best indicator of foraging success at any given time is the intensity of flight activity at the hive entrance.

Successful foragers collecting food near the hive (within approximately 100 yards) perform "round" dances on the surfaces of brood combs. Bees foraging at greater distances perform "wag-tail" dances in which directional information is coded into the

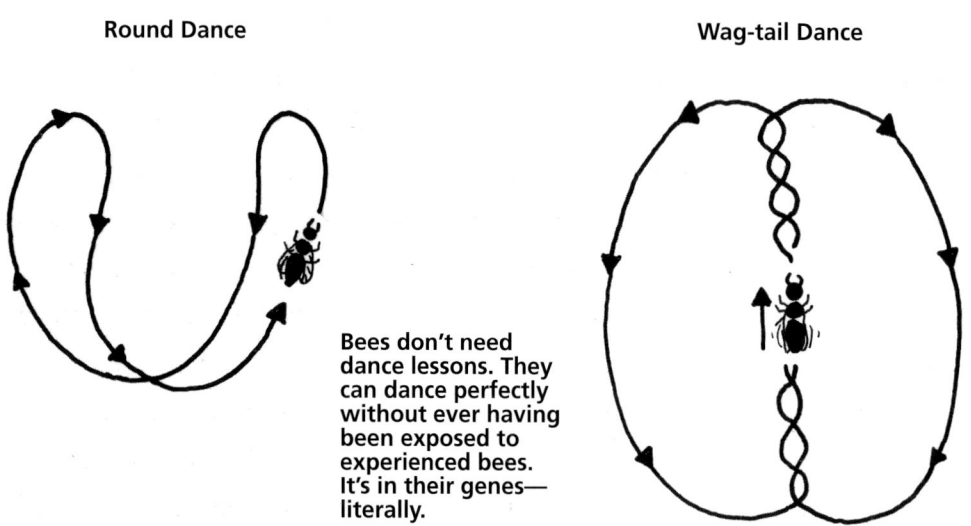

**Round Dance**

**Wag-tail Dance**

Bees don't need dance lessons. They can dance perfectly without ever having been exposed to experienced bees. It's in their genes—literally.

dance. Dancing bees also produce tiny, rapid buzzing sounds in variable numbers that are highly correlated with distances from the hive to the food.

## Feeding Larvae

Worker bees become nurse bees when they are a few days old. Their brood-food glands start secreting nutritious liquid brood food that is fed to developing larvae. Up to three days old, the tiny larvae float in a C-shaped position on a bed of the liquid food at the bottom of the cells. Nurse bees wander around in total darkness until they encounter cells where larval pheromones stimulate them to enter the cells and deposit food. Once again, these behaviors are instinctive, not learned. Objectively speaking, nurse bees don't even know what *nursing* means or that they are nursing. They never actually see a larva or the cell. They simply respond appropriately when stimulated.

## Food Processing and Storage

Foraging bees deliver pollen to the hive in the form of colorful pellets adhering to the pollen baskets on their hind legs. To unload, a pollen-laden bee backs partially into an open cell and dislodges the pellets into the open cell. These pellets are immediately compacted into the cell bottoms by other bees. They incorporate honey and other salivary secretions. After processing and storage are completed, the cells containing pollen are typically about two-thirds full and not capped. Complex chemical changes—similar to fermentation—occur during the next few days, converting the raw pollen into a gummy mass known as bee bread. Nurse bees eat bee bread, which stimulates the production of secretions they'll feed to larvae and to the queen. Pollen supplies all nutrients (other than the carbohydrates supplied by

honey) required in the diet of honey bees. Just like other animals, bees need proper nutrition for good health.

*House bees* (bees that haven't reached the foraging age) unload nectar-laden foragers returning from foraging trips by sucking up nectar droplets presented on the mouthparts. The house bees then unload the nectar into open comb cells. Nectar foragers also may unload directly into comb cells. Fresh nectar contains an average of about 70 percent water; it will literally drip from the open honeycomb cells when the combs are being examined. Beekeepers sometimes monitor the honey flow by shaking the comb in a horizontal position with the open cells pointed downward. They are delighted when nectar rains copiously from the comb.

Nectar becomes honey via two processes: the evaporation of excess moisture and the addition of enzymes that break complex sugar molecules into simple sugars. Bees reduce the water content in nectar by spreading thin layers of nectar on their proboscis and in cells and then fanning it. Honey, by definition, contains around 14 to 18 percent water. A high concentration of sugar-tolerant natural yeasts typically

**To unload the pollen load on her legs, a worker bee will find an empty cell, back into it, and drop off her two pellets. Freshly collected pollen is packed into cells until they are about two-thirds full.**

found in honey will cause fermentation and spoilage if moisture exceeds around 18.6 percent. The solubility of simple sugars compared to complex sugars is greater in the aqueous solution we know as honey. Increased solubility enables the bees to pack more calories into the limited honey storage space. Storage capacity can be the difference between death and survival of the colony during the long winter months when the bees are confined. As soon as the bees convert nectar into "ripe" honey, they seal the

The upper part of this brood comb shows cells capped with beeswax that contain honey for long-term storage. Open cells below contain stored pollen.

cells with caps made of beeswax. These caps preserve honey perfectly during extended storage times, enabling bees to survive long periods on reserve food inside the hive.

Here you see sparkling fresh nectar interspersed with freshly collected pollen.

Honey is stored immediately above the brood nest. As the combs become full, the bees' instinctive response is to expand the storage upward. Beekeepers try to anticipate the need for space and add additional chambers (supers) of comb on top of the colony as needed. Failure to anticipate adequate nectar and honey storage results in lost honey production and excessive honey storage in the brood nest, which reduces the comb space available for rearing brood.

## Drone Activities

Drones do not participate in any maintenance activities inside the hive. They tend to congregate on the outer sides of the brood-rearing area. They feed themselves each day just before they depart on mating flights during the afternoon. Drones have been maligned in popular literature as being

This drone is revving his engines as he takes off from the hive entrance on an afternoon flight.

lazy and irresponsible. This isn't really fair. Drones don't have the body structures that would make it possible for them to do work inside or outside the hive. They don't have wax glands, they don't have stingers, and their proboscises are so short that they couldn't possibly reach nectar deep inside flowers. But they are very efficient sperm machines. Isn't that enough? Do we criticize human males as being lazy because they don't nurse babies due to the absence of functional mammary glands?

## Division of Labor

Writers have described the natural process of different bees doing different things at different ages as "division of labor." In the first place, bees' activities are not labor. Labor is for humans, not insects. In reality, bees have no awareness of this concept and no grasp of the impact of their current activities on colony welfare or future events. They just respond to their immediate environmental stimuli. Their activities, although exquisitely coordinated and appropriately functional, are not the result of intelligence and cognition. Fortunately, you can be a skilled hobby beekeeper without concerning yourself with these philosophical matters. If you think bees are intelligent, then they are—but you have an obligation to equate them with other insects to be fair. Are there intelligent cockroaches, termites, or mosquitoes? You should take a moment to think deeply about these issues. You have the opportunity to control your bees if you fully understand that they are unthinking, reflexive critters. This greatly simplifies beekeeping operations. All you have to do is understand their predictable behavior and perform your beekeeping operations accordingly.

Let's explore the inner world of the honey bee regarding the *what*, *when*, *why*, and *how* aspects of their activities inside the hive. It's well known that they perform a succession of activities—popularly described as "duties"—according to their age. As honey bees age, the physiology and development of their body parts change, enabling them to perform different behaviors. For example, newly emerged adult bees—sometimes referred to as "baby" bees even though bees come out of the comb as fully developed adults—have stingers, but they are too soft and flexible during the first few hours to penetrate skin. In addition, their venom glands are not "loaded" with venom at that tender age. These young bees will not and cannot behave defensively by trying to sting you.

Although a given bee typically engages in several different activities, such as secreting beeswax or feeding larvae with brood food, before becoming a forager, the system is amazingly flexible in response to the colony needs. For example, experiments in which researchers removed all of the older foraging bees from the colony dramatically accelerated the aging process of young bees. Some of them became foragers a week or so earlier than they would have under normal conditions. In a short time, the population adjusted to the changes, and the delicate balance of activities inside the hive was restored. The sequence of activities for a given bee is not rigid, and some bees skip certain behaviors that would be expected for their age. Genetic influences undoubtedly affect their behavioral tendencies.

How do bees know where to go and what to do in the dark environment of the hive? The key is that they move around a lot, thus exposing them to situations that "fit" their behavioral tendencies at that moment. A bee with engorged brood-food glands would be "primed" behaviorally to sense a hungry larva in the cell near her and respond reflexively by feeding it. In the human world, the equivalent behavior is a hungry snack-food fan passing by a fragrant popcorn stand and responding reflexively by buying the biggest size possible. (Bees don't have a monopoly on instinctive behavior.) This simple mechanism of bees' moving around until the environmental stimuli trigger their appropriate behavior is perhaps the best way to understand the complex behavioral events inside the hive.

A breathtakingly beautiful field of lavender could be thought of as a Disneyland for foraging bees. Planting nectar and pollen plants like lavender in urban environments can greatly benefit bees.

# CHAPTER 7

# Activity Outside the Hive

What happens inside the hive is the domain of the beekeeper and is the source of endless pleasure for those who appreciate playing and working with bees. What happens outside the hive involves the whole world—literally. Honey bees have a great impact on the welfare of people and animals worldwide. If we could somehow remove honey bees from the earth for a year, then we could truly appreciate their contributions to food production and plant survival. Without question, there would be catastrophic consequences. Since we can't perform this grandiose experiment, let's at least look at some of the activities of bees outside the hive to develop a better understanding of their value to humankind.

# Pollination

Bees play a fundamental role in food production. About one-third of the food we eat, at least in the United States, can't be produced without pollination by honey bees. Fruits, vegetables, berries, some fiber crops, domestic animal feed, and oil seed crops would be in extremely short supply without honey bee pollination. Can you imagine the impact on our food supply and diet if honey bees weren't available for pollination? Without them, the human diet would consist mostly of grains and fish.

# Plants that Yield Pollen and Nectar

Many species of plants yield nectar that is attractive to honey bees. Nectar is produced in *nectaries*, the specialized glands that secrete a solution containing various kinds of sugars. Most nectaries are inside flowers (*floral*), but they may be anywhere on the plant surface (*extrafloral*). A few plant species (cotton is a good example) have both types of nectaries. Don't be surprised if you observe bees busily collecting nectar on petals, sepals, and petioles; between bracts; under leaves; or in other unexpected locations.

Nectar is a watery solution of various sugars, primarily glucose, fructose, and sucrose (table sugar). Typical nectars contain an average of about 30 to 40 percent sugar concentration at the time they are collected from the flower. Other constituents, usually in very small amounts, are proteins, amino acids, lipids, organic acids, dextrins, minerals, antioxidants, and enzymes. Volatile chemicals, perceived as fragrances, are important ingredients, too. They not only attract bees to the nectar sources but also typically contribute distinctive flavors to the finished product—delicious honey. Although nectar is complex in chemical composition, most of the ingredients other than sugars do not contribute significantly to bee, or human, nutrition.

# The Foraging Process

Plant species that yield collectable nectar and pollen typically are not distributed homogeneously in the environment. They are found in various patterns, ranging from large areas of monocultured fields in contemporary agricultural areas to widely dispersed patches of flowers in natural settings. Recruitment of new foragers using dances tends to distribute foragers according to the relative profitability of plant species within the area.

In addition to recruitment dances, there is another means of communication that adjusts the foraging activity of the experienced forager bees. The entire experienced forager population doesn't automatically leave the colony *en masse* each day. Limited numbers of scout bees go first. When they return from successful forays, they signal the status of food availability by sharing samples of nectar with the previously experienced foragers that are on "standby." Nectar-sharing, combined with flower fragrance on the bodies of the returning foragers, signals the availability of pollen and nectar. Experienced foragers already know the area, so they take off to resume foraging in their respective familiar areas.

Individual foragers return to the same foraging areas time after time—as long as the profitability of the source exceeds the cost of energy and time to fly there and to gather a load. This is known as *area fidelity*. Each bee also typically forages on one species of flower. This is known as *species fidelity*. Area and species fidelity by individual bees is characteristic foraging behavior. If you monitor a given honey bee collecting nectar and pollen in an area where several plant species attractive to bees are intermixed, she will meticulously go from flower to flower of her "favorite" species. She ignores other species that are simultaneously yielding profitable loads of nectar and pollen to other bees—even to sisters from the same hive. A few foragers do not follow these "rules," however. Some return to the hive carrying pollen loads of mixed colors—evidence that they collected pollen from more than one species on that trip. Some bees forage over wide areas and are not

**The production of flower seed requires pollination by honey bees. As evidenced by this colorful landscape, it's big business.**

faithful to a given foraging area. These wandering foragers apparently provide an ongoing scouting function.

During research studies, many thousands of bees were observed in diverse habitats where apiaries were randomly distributed over large areas. The resulting foraging distribution maps clearly revealed important parameters of the foraging process, some of which I've already mentioned. First, the distribution of foraging bees is not simply a symmetrical circle around the apiary. Competition from nearby apiaries changed the distribution pattern such that very few bees from either apiary foraged beyond the halfway point between the apiaries. The foraging population in each colony preferred a diversity of plant species at various distances as opposed to foraging at the nearest source of one productive species. In summary, individual bees forage no farther than necessary to reach their "favorite" flower species. If the nearest source of an attractive species is located 4 miles from the hive, they will go there. Along the way, they will ignore and bypass other rewarding foraging areas and different species simultaneously frequented by sister bees from the same colony. This behavior dramatically demonstrates species fidelity by individual foragers. Subgroups of foragers in a given colony, each faithful to their respective species, collectively gather pollen from many species at the same time. Eclectic foraging on many species ensures adequate nutrition because the nutritional value of pollen varies significantly between plant species.

## MAGNETIC ATTRACTION

The directional distribution and foraging distances of foragers have been studied extensively. One novel method was especially productive. Bees were captured while foraging—thereby documenting the flower source and the area—and tiny metal identification tags were glued to their abdomens. The tags were responsive to magnetic attraction. When tagged bees return to their respective hives, entry was contingent upon passing beneath banks of horseshoe magnets placed strategically over the hive entrances. The bees were unceremoniously yanked off their feet (imagine the little bee profanities). As they struggled to get free the tags were pulled off—without harming the bees—and recovered by the magnets. Consequently, the distance and direction to the foraging area for each bee was documented.

**Here's a great example of the multicolored pollen that bees can collect from a variety of flowers.**

## Nectar Foraging

Bees will not collect nectar if the sugar concentration is too low—less than around 5 percent. The quantity of nectar per flower varies enormously. For most plant species, bees must visit hundreds of flowers to get a full load. Amazingly, a full load of around 60 microliters weighs about two-thirds of the bee's body weight when the honey stomach is empty. Inexperienced foragers quickly learn to associate flower colors and shapes—yes, they can see colors and shapes—with the food reward. These visual cues help them find flowers quickly. Their extended tongues are long enough to reach the nectar deep inside flowers. Nectar is then temporarily stored in their honey stomachs for transport back to the hive. The bees unload quickly and immediately resume foraging. Bees can make many trips daily. They continue to forage as long as nectar is available. To make a pound of honey, the bees collectively have to fly a distance estimated to be at least twice around the world—around 50,000 miles.

There is a daily temporal pattern of nectar and pollen availability in each plant species. Some species produce early in the morning and may cease within several hours. Others may be productive throughout the day. When a species supports foraging in the morning only, then bees with fidelity to this species get the rest of the day free to do whatever bees do when they get free time, milling around inside the hive or assuming a rest position—head upward and snuggling side-by-side with other idle workers. This behavior appears to be the bees' equivalent of sleep, but since they don't have eyelids, we can't really know when bees are sleeping.

Under ideal conditions, plants can secrete large amounts of nectar, stimulating peak foraging activity. The loud hum of flying bees

**On a lazy, warm afternoon, it's fun to lie near your hive (not too close) to watch bees coming and going like planes at an airport.**

at the apiary—thousands of bees coming and going—may be disconcerting to the uninitiated, but it is music to the ears of a beekeeper. Sometimes the fragrance

## ALWAYS ON TIME

Bees have a very accurate sense of time that enables them to synchronize foraging activities with nectar and pollen availability. Experimentally, bees can be trained to collect artificial nectar at a feeder at different times of the day—let's say nine o'clock, twelve o'clock, and three o'clock. Remove the feeder, and the foragers will still arrive at these times.

of nectar being processed inside the hives fills the air with a lovely aroma similar to the fragrance of the flower source. Some nectars produce unpleasant odors at the hive entrance during the curing process but, contrary to expectation, result in delightfully flavored honey. When all conditions are ideal (good weather, long days, intense nectar secretion, and very populous colonies) bees can collect enormous quantities of nectar—perhaps around 6 pounds or more in one day—yielding around 2 to 3 pounds of honey per day.

## Pollen Foraging

The importance of pollen to the health and vigor of the honey bee colony cannot be overstated. Bees need a balanced diet. Honey satisfies the bees' carbohydrate requirements, while all of the other nutrients—minerals, proteins, vitamins, and fatty substances—are derived from pollen. Nurse bees consume large amounts of pollen, converting it into nutritious secretions that are fed to developing larvae. During an entire year, a typical bee colony gathers and consumes about 77 pounds of pollen.

The pollen that bees collect is composed of larger, heavier grains (compared to airborne pollen) that tend to be sticky. Pollen plants have a negative electrical charge that perfectly complements the positive charge that builds up in the bee as she flies. When she contacts the flower, pollen grains are attracted to her body, similar to the attraction of iron filings to a magnet. Pollen is released from a plant's anthers by *dehiscence*—the splitting- or bursting-open process that releases pollen grains. Plant species may dehisce at different times during the day. Bees learn to coordinate their foraging visits with the time of dehiscence for each plant species.

Most of the external structures on the bee's body are functional during pollen collection. Using the hairs that cover their bodies, they brush against the flower structures as they collect nectar and/or pollen. Some bees collect pollen only. Others collect both nectar and pollen on the same trip when these substances are available on the same plant species. Inexperienced foragers quickly learn the most efficient movements to dislodge pollen grains from the anthers. This frequently involves the use of their tongues and mandibles in licking and scraping the anthers, which slightly moistens the pollen. After becoming coated in pollen grains, the bees meticulously

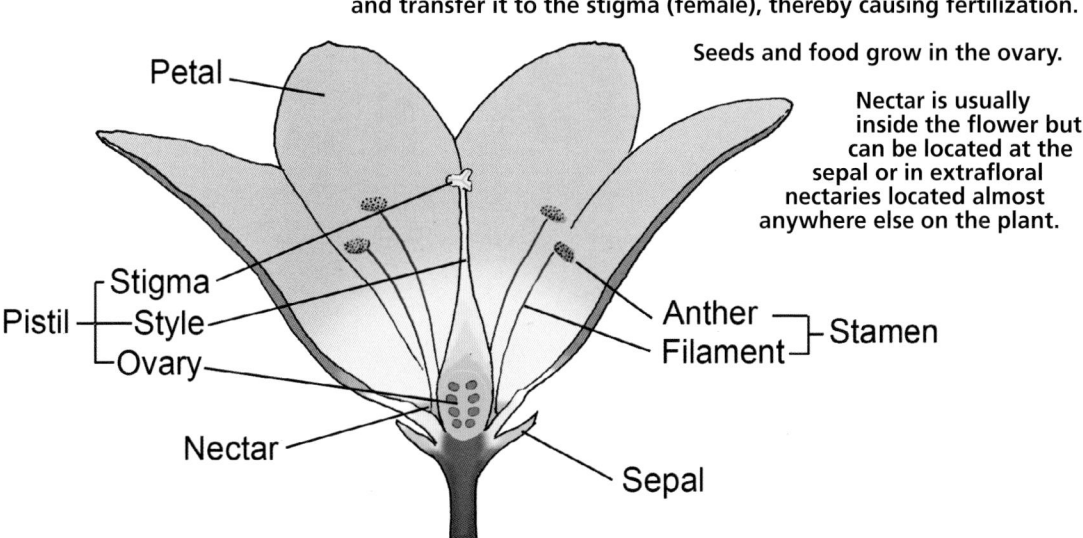

Bees accidentally brush against pollen released by the anther (male) and transfer it to the stigma (female), thereby causing fertilization.

Seeds and food grow in the ovary.

Nectar is usually inside the flower but can be located at the sepal or in extrafloral nectaries located almost anywhere else on the plant.

Petal

Pistil — Stigma / Style / Ovary

Anther / Filament — Stamen

Nectar

Sepal

**Pollen grains adhere to the bee's hairs, influenced by opposite electrical charges. Unusually large grains from a pumpkin flower are shown. Notice how the extended hairy tongue (proboscis) wicks the nectar upward so it can be sucked into the mouth.**

brush pollen from the body hairs with the comblike hairs on their legs. This process gradually transfers the pollen to the hind legs, where it accumulates as two "pellets," adhering to the outer surfaces of the pollen baskets on their hind legs.

Transferring pollen from the body to the pollen basket involves a remarkable combination of synchronized combing movements by all of a bee's legs— choreography that almost defies description. Sometimes the bee interrupts her collection and rests on the flower to comb and brush herself. More commonly, she hovers around the flower, performing these brushing maneuvers while flying—an incredible display of coordination. Fortunately for the plants, bees aren't 100 percent efficient at transferring the pollen to the pollen baskets. Thousands of pollen grains may still remain on their bodies even after they finish grooming. Bees leave enough pollen behind, depositing it accidentally on female flower structures, to ensure effective pollination.

Ten pollen-foraging trips per day is par for the typical pollen-forager. When pollen is abundant, a bee can gather a full load in as little as ten minutes by visiting several dozen flowers. As available pollen diminishes, a bee may have to spend an hour or

## POLLEN 101

Pollen in the plant world is the equivalent of sperm in the animal world. Fertilization and growth of seeds depend upon the transfer of pollen from the male flower parts (*anthers*) to the receptive female parts (*stigmas*). For many species, such as grains and nuts, pollination occurs by airborne pollen that is produced in great quantity and characterized by very small, lightweight pollen grains. Airborne pollen causes most human allergies.

This pollen-collecting bee is homing in on Rock Purslane, a plant native to Chile. The bright magenta flowers are highly attractive to bees.

more visiting hundreds of flowers to get a full load. When all factors are favorable, bees from a strong colony can collect many thousands of loads a day. Incidentally, when a beekeeper describes his colony as *strong*, he really means *populous*—even though bees as individuals seem to have Herculean strength, flying with loads nearly as heavy as their body weight.

Over millions of years, plants and bees have interacted in ways that have led to complementary changes in flower structure and bee behavior, leaving both better adapted to reproduction and long-term survival. Some bee species are specialists. For example, squash bees specialize in pollinating cucurbit plants (squash, cucumbers, and others). Honey bees evolved in a different direction, developing appropriate tastes and behaviors that underlie their preference for collecting nectar and pollen from a diversity of plant species. The "one–size-fits-all" pollination capability of honey bees is a major reason they are used commercially to pollinate agricultural crops.

## Water Foraging

Bee colonies require small quantities of water—up to around 7 ounces per day—but the water they collect is vital to their survival. During periods of hot weather, bees evaporate tiny droplets of water in the hive to control the internal colony temperature. Maintenance of internal colony humidity is important to developing larvae. In addition, nurse bees use water to reconstitute honey to a nectarlike consistency when feeding larvae. Bees also drink water to sustain their own bodily functions. Providing a water source is not necessary for the hobby beekeeper in most areas because the moisture in the nectar that the bees collect satisfies most of their water requirements. In addition, there typically are many water sources, such as leaky faucets, within flight range in urban environments.

Water foragers tend to forage at the water source nearest to their colony. Minerals, salts, gases, organic compounds from organisms in the water, and other unknown elements influence the bees' preference of water sources. Only the bees know the secret ingredients that determine their choices; otherwise, beekeepers would be able to create super-attractive watering

A lone bee sucks up a load of water needed back at the hive.

It's not fancy, but bees love to drink water from an inclined board where water drips.

holes with special flavors to lure bees away from swimming-pool decks, drinking fountains, and birdbaths, where they are sometimes perceived as a problem by nonbeekeepers. People don't realize that water collectors are not a sting threat, because stinging behavior normally happens only near colonies that have been disturbed. Beekeepers do sometimes try to provide water sources in an attempt to discourage water foragers from sources where they are unwelcome guests. Probably the most successful homemade water-feeder design is an inclined board mounted under a slowly dripping water source. This may not work well at the outset, but the surface seems more attractive to bees as it becomes covered with algae over time.

It seems impossible. Somehow bees collect sticky propolis and turn it into smooth, gleaming jewels in their pollen baskets. Propolis is used as caulking to seal cracks inside the hive walls.

## Propolis Foraging

Propolis, sometimes called *bee glue*, is a gummy, resinous, sticky, thermoplastic, brown exudate from the buds of trees that varies enormously in composition and color because it is collected from tree buds and sap from a great variety of botanical sources. It's usually dark brown. Bees use it as a caulking material to seal small cracks and crevices inside the hive, especially at the joints between chambers, making it difficult to separate hive chambers that are glued tightly together. In warm weather, propolis is sticky and pliable. In cold weather, it's hard and brittle.

A few bees in each colony collect propolis during warm weather when it is pliable enough to manipulate. This natural bee glue is so sticky that other bees need to assist the propolis forager during the unloading process. It's amazing how they can manipulate it without becoming hopelessly entangled and stuck together, but they seem to manage just fine. Beekeepers, however, aren't as lucky. It doesn't take long for new hive equipment—so easy to separate at first—to become glued together. To remove propolis from your hands or tools, it's helpful to use hand cleaners manufactured for cleaning a mechanic's greasy hands in the auto shop.

The irregular brown bead of propolis on a frame's contact point makes frames stick together so a hive tool is necessary to remove frames of comb.

Sometimes a bee really has to dive deeply into a flower to reach the nectar.

## Weather and Foraging

Weather can profoundly affect foraging activity. Foraging flights can start at daybreak if the temperature is warm enough, although cool morning temperatures may delay foraging until later in the day. Foraging activity is most active when the temperature is between around 60–100 degrees Fahrenheit. Bees and most plants function well in this temperature range. Windy days can slow or even stop bee flight in the field whenever the wind velocity exceeds the bees' maximum flight speed of about 14 miles per hour. When the wind is blowing, bees fly much lower to the ground, where the wind velocity is less. This can result in many casualties when bees are crossing roads during high-speed traffic conditions.

Rain can severely affect foraging activity in several ways, as it may (a) dilute nectar, (b) wash it out of the flowers, or (c) reduce its sugar concentration to the point that bees won't collect it. Foraging activity is also affected by dark skies and electrical disturbances. Just minutes before the onset of a storm, somehow the bees sense trouble brewing, and there is a huge traffic rush of foragers returning to the hive to escape the potential punishment they associate with the storm. (One might suppose that an in-flight collision with a big drop of rain would be a punishing experience for a bee.)

Weather can also affect nectar secretion. Optimum nectar secretion in some plants requires hot days in combination with cool nights. Some plants require abundant rainfall while others do well under extremely arid conditions. The dynamics of nectar secretion are known but not entirely predictable—similar to weather forecasts. Beekeepers have to learn the idiosyncrasies of nectar and pollen plants in their areas in order to manage colonies efficiently.

## Hive-Entrance Activities

You can observe many activities at the hive entrance, which is basically a runway where bees land and take off. It's fun to watch heavily laden returning foragers, many with brightly colored loads of pollen, coming in for a landing. Guard bees position themselves near the entrance to defend the nest from intruders. On hot days, you'll see dozens of fanning bees on the entrance. Bees on one side may be heading away from the hive, while bees on the other side are heading toward the hive. Why the opposition? There is an intake side and an exhaust side. Gently suspend a feather at the entrance and compare the direction of air movement as you shift the feather from one side to the other. If you puff smoke into the intake side, the fanning bees will reverse the circulation to exhaust the smoke—the intake becomes the exhaust and vice versa. If you spend time watching the current events at the hive entrance, be careful not to block the flight path or exhale in the direction of the guard bees, or the bees may show their displeasure at your presence.

**This guard bee, standing at attention at the hive entrance with sensitive antennae pointing up, is ready to tackle any intruder, regardless of size.**

# CHAPTER 8

# Colony Defense & Sting Prevention

The fear of stings is probably the greatest deterrent to hobby beekeeping. Understanding defensive behavior and skillfully applying sting-prevention strategies essentially determines (a) whether or not you will choose to make beekeeping your hobby, (b) how much enjoyment you will derive from keeping bees, and (c) how successfully you, as a beekeeper, will interact with family, neighbors, and pets.

# Bee Stings

An occasional bee sting comes with the territory, comparable to the small risks associated with most pets. Cats scratch, dogs bite, horses kick, and birds peck—just to name a few. There are ways to neutralize a stinger so that getting stung really doesn't amount to much, and certainly not enough to deter you from enjoying your new hobby. But before you start keeping bees, you should see your doctor for an allergy test. You don't want to discover you are allergic to the venom released in bee stings while standing amidst a group of hives with bees surrounding you (see Reactions to Stings).

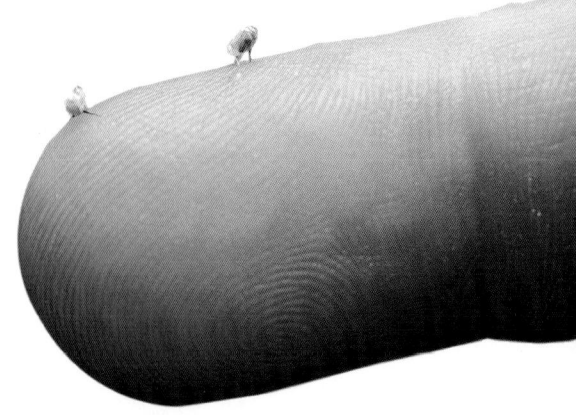

**These stingers, only halfway inserted here, will pull themselves in completely within several seconds, delivering venom as they penetrate.**

Almost everyone has been stung by an insect, and most people believe that the sting came from a bee. However, most sting incidents involve wasps, which are very much *not* bees. How can you know if you have been stung by a wasp or a honey bee? That's easy. A honey bee can sting only once because the stinger pulls free from its body and *thus will be easily visible protruding from your skin*. All other species of stinging insects, including wasps, can sting multiple times, and they don't leave the stinger behind as evidence.

During routine day-to-day activities, people rarely encounter bee hives—the only places where there is a significant sting risk. Honey bees simply are not inclined to sting when they are away from the hive. But accidents happen: you step barefoot on the little creature that is harmlessly collecting food from flowers, or maybe you

**Yellowjackets, which are wasps, are similar in size to honey bees. Despite the color differences, many people think yellowjackets are bees.**

collide with or swat at a flying bee, thereby forcing an involuntary sting. In the case of honey bees, these stings are isolated accidents. By contrast, yellowjackets—wasps that are commonly referred to as *meat bees*—behave differently. They are uninvited guests that frequently visit your picnic table or pester you almost anywhere outdoors when they detect the aroma of meat. (Remember, honey bees are not attracted to meat.) If you don't monitor every bite of your sandwich very carefully, you may accidentally eat one alive—not advisable. They also can enter and get trapped inside your drink container, providing quite a surprise at your next sip.

Honey bees are defensive, not aggressive. Defensive behavior is necessary to the survival of the colony. If access to the combs was easy and painless, enemies would eat the bees, their young, and the stored honey and pollen.

Hover flies mimic the appearance and behavior of bees, fooling enough enemies to receive the same respect as a bee with a stinger.

*Defensive behavior happens only when you are very close to the hive.* Bees foraging on flowers or collecting water behave as if they are totally defenseless. They will fly away in response to the slightest disturbance. A bee that has stung dies within a few hours. The colony benefits only if this sacrifice is made in defense of the colony. Therefore, stinging away from the immediate hive area is not normal for European honey bees.

The act of stinging near the hive is a reflexive response stimulated by your behavior, so learning how to act around your bees is important. Reflexive responses happen automatically when certain nerves are stimulated; there is no thought process to control reflexes. Remember when your doctor gently tapped your knee and you kicked out your leg uncontrollably? A few individuals in the colony population, commonly called *guard bees*, are primed behaviorally to respond reflexively to the actions of animals or people that disturb the colony. A loaded gun fires when the trigger is pulled. When you trigger a guard bee's nervous system, the bee simply reacts by stinging. Again, the act of stinging is not motivated by any kind of thought process. The good news is that this defensive behavior is predictable and controllable to a very great degree.

## Why Do Bees Sting?

The tendency to sting is greatly affected by a bee's genetic makeup—similar to the tendencies of various dog breeds to bark or bite. Bee-breeding research confirms that it is possible to produce bee colonies that show little inclination to sting. Defensive behavior varies greatly between colonies. Several factors underlie and control the defensive responses of guard bees at any given time. *Bees sense and are excited by vibrations, motion, dark colors, exhaled breath, and body odors.* Some clothing materials also exacerbate stinging. For example, suede leather (a) has a texture that allows the bees to grasp it easily, (b) has physical properties that permit easy penetration by the stinger, and (c) has residual animal odors. Never wear suede clothing near the hive. Remember also that the natural tendency of European bee colonies is to defend the area near the hive only. If you don't like the attention, you can quickly escape by simply leaving the hive area.

Other factors also affect the degree of defensiveness. Although very populous colonies are desirable and necessary to maximize honey production, they are likely to be more defensive than small colonies. The time of day is an important consideration, too. Bees become more defensive just before sundown, and this behavior persists until dawn. Chronic molestation by animals, such as skunks, ants, or other pests, can be stressful to the colony and greatly increase your chances of getting stung. Colonies are least defensive from mid-morning until mid-afternoon, or any time when flight activity of foragers at the hive entrance is greatest. They are especially docile when nectar and pollen are abundant.

## How a Stinger Works

A flying bee grasps the surface—maybe your skin—with her tarsal claws to position herself. Then she bends her abdomen downward and stabs with the sharp, needlelike tip of her stinger. This "needle" isn't just a hollow tube with a smooth exterior surface like the familiar syringe used for injecting medications. It is actually composed of three parallel parts that fit snugly together with a tongue-and-groove arrangement, forming a tubelike canal in the core that is called the *poison canal*. Two of these parts—called *lancets*—have serrated and barbed edges, and the third part (*stylus*) anchors and guides the two lancets, which can slide back and forth. Once the lancets penetrate the skin, their barbs catch and hold, similar to the action of a fishhook—easy penetration, but difficult withdrawal. Once the stinger penetrates and the lancets catch, the entire stinger is pulled out from the tip of her abdomen as she leaves the sting site. After the bee flies away, she continues the next phase of defense—intense *sham attack flight* in which she flies around you and repeatedly attempts to sting even though she is now stingless. This behavior probably attracts and stimulates other defensive bees to join the attack.

Once separated from the bee, the stinger spontaneously springs into action. It has its own nerves and muscles that function automatically. The two lancets begin to thrust alternately, each penetrating deeper into the flesh and holding position, literally pulling the stinger shaft deeper into the wound until it is fully inserted. You can see the stinger protruding from your skin; its pumping movements are visible even without magnification. The stinger then delivers its venom in tiny squirts through the stinger shaft. *Most of the venom is delivered during the first few seconds.*

A few minutes after the stinger is fully embedded, a localized reaction becomes obvious.

## Defusing the Sting

Now that you know how a stinger works, you can do something to minimize its effects. *You must take action very quickly to reduce the amount of injected venom.* Obviously, severing the connection between the visible part of the sting (where venom is stored) and the delivery tube (stinger shaft) would terminate the injection. Common advice from beekeepers is to scrape away the stinger with the edge of a credit card. If you are not wearing bee gloves, you can just scrape the stinger out with your fingernail. *Any technique that destroys the venom sac within two seconds is good.* Speed trumps technique. If you terminate venom delivery almost instantly after being stung, the effects are greatly minimized. You aren't actually receiving a normal sting—just a tiny sample. Delayed stinger removal is a fruitless exercise.

A freshly deposited stinger releases an instant "odorous flash" of powerful alarm pheromones at the sting site. Pheromonal stimulation combined with the post-sting buzzing of a stinger donor around the target excites other nearby defensive bees. *The odds of getting stung again dramatically increase after the first sting.* As soon as possible, move a few steps away from the hive (out of the defensive area) and wash the sting site with water—nothing more. Cool water feels good and washes away the pheromone odor.

It is a comfort to know that the pain lasts for only about a minute. If you want to minimize the pain, you can hold an ice cube on the sting site for a minute or two. (For some people, a brief episode of profanity also helps.) Now you can return to the hive, smoke it well, and start afresh. Applying topical medications or lotions at this time may actually increase the risk of receiving additional stings because of the associated odors.

## Reactions to Stings

The intensity of the pain depends upon the concentration of nerve endings at the sting site. A sting on the tip of the nose is memorable, but a sting on the arm or leg may be barely noticeable. Psychological pain may exceed physical pain, especially for someone who expects the worst possible consequences. The expectation of pain can magnify the experience beyond a rational reaction. An experienced beekeeper knows what to expect and calmly deals with a sting as if it were a mosquito bite.

Work with a beekeeper pal in the apiary, especially for the first few months, until you gain confidence and your reaction to a sting can be established. Don't judge your sensitivity by your response to the first sting. An initial exposure is required to activate hypersensitivity for the 1 percent of people who are inclined to develop hypersensitivity. You won't know your sensitivity status until days, months, or years later, when you get the second sting. The most common and obvious symptoms of a serious reaction—*anaphylaxis*—are difficulty in breathing caused by swelling in the throat, nausea, weakness caused by a sudden drop in blood pressure, dizziness, and rash. These symptoms will develop quickly within the first few minutes after the sting and must be treated quickly with an injection of epinephrine (synthetic adrenalin). Epinephrine is available by prescription from your doctor. A genuine anaphylactic

## CAUTION: LIFE AND DEATH

Before you commit to beekeeping, you should consult your doctor about hypersensitivity to honey bee stings. Only about 1 percent of people are hypersensitive, but if you're one of them, this little visit to your doctor could be the difference between life and death.

reaction develops fast and must be treated immediately. Delaying treatment to see if the symptoms will diminish can be fatal. Check the expiration date on the epinephrine. The solution should be clear like water, not discolored. Avoid exposing it to heat during storage. Besides epinephrine, your safety net if you get a serious reaction to a sting is your cell phone. Keep it handy. **If you begin to have an allergic reaction and do not have epinephrine on hand, call 9-1-1 immediately.**

The overwhelming odds are, however, that you are normal—at least in terms of your reactions to a sting. Serious reactions are rare. Overall, *honey bee stings are easy to prevent and can be essentially neutralized by quick action.* Many people mistakenly believe that they are allergic to bee stings whenever they experience normal reactions— localized swelling, inflammation at the sting site, and itching for a day or so. These are typical reactions to bee venom. Beekeepers who receive the occasional sting normally become increasingly tolerant. In fact, many victims of rheumatism or arthritis benefit from bee-venom injections, a type of treatment known as *apitherapy* (see Bee Venom in Chapter 10). If you are concerned about the slim possibility of hypersensitivity, a physician can test you for an allergic reaction before you commit to hobby beekeeping.

## How to Behave Near the Hive to Prevent Stings

Proper etiquette near the hive is essential. By controlling your behavior in certain ways, you can avoid or minimize those "red-flag" mistakes that cause bees to sting. First, do not breathe on the bees. Exhaled breath contains carbon dioxide as well as other odors that can instantly trigger a stinging response. When possible, stand downwind from the open hive. Momentarily hold your breath when your face is near the combs covered with bees. When you need to breathe, exhale downwind so your odors blow away from the hive. Breath odors are to bees what a waving red cape is to a bull. Bees react instantly. Exhaled carbon dioxide is the common denominator for living animals, many of which were bee predators during the bees' long evolutionary history. It's not surprising that bees developed extreme sensitivity to carbon dioxide. If you ever see a beekeeper blowing the bees on the comb to make them move over to enhance visibility of the cell contents, move farther away from the hive and watch the exciting show from a distance.

Defensive bees are very sensitive to motion. You should make all of your movements in slow motion. It's also important to minimize your hand movements directly over the open hive. When removing frames of comb, reach around the sides of the hive—out of sight of the bees that are perched on the tops of comb frames, watching you. A simple demonstration can show why slow motion is essential.

When an alert guard bee is looking at you with her front pair of legs held in this position, she means business. A puff of smoke will defuse this defensive threat.

Caution: *wear bee gloves for this demonstration*. Open the hive, using minimal smoke. Move your hand very slowly, about 6–10 inches over the frame top bars while watching the bees that are watching you. They won't react to your slow motion on the first pass. On the second pass, move your hand a little faster, and you should see a few bees turn their heads to follow your movement. On the third pass, move your hand faster past the bees. Some bees usually react instantly—flying and trying to sting. You get the idea. Make it a habit to move slowly around and near the open hive.

## Alarm Pheromones from Injured Bees

Be careful to protect bees from accidentally being crushed during comb manipulations. A crushed bee releases alarm pheromones from its stinger. These pheromones stimulate stinging behavior by other nearby bees, thereby increasing the chances for you to get stung. Subsequent stings would release additional alarm pheromones. A chain reaction of stinging can ensue quickly—within several seconds. Use the following directions to minimize the risk of injuring bees.

During hive inspections, you have to remove the frames for examination. Removing the first frame is challenging, especially when the frames are wedged tightly together and glued with propolis. Plan ahead to make the job easier. First, after comb construction in the new frames is completed, routinely use nine frames in a ten-frame chamber. Leave a generous bee space between the outer two frames and the chamber walls. Position the nine frames together, equally spaced, in the middle of the chamber. Thereafter, during routine hive examinations, always remove one of the outer frames (next to a chamber wall where there is more working space) first to minimize the risk of accidentally crushing bees—especially the queen. To make enough working space for easy frame removal, slide the frames, perhaps three at a time, toward the opposite chamber wall. To do this, insert your hive tool about 2 inches between the frames, where they make contact on each end, and use the leverage to pry the frames apart and slide them.

Once the working space is created, cautiously pull the comb straight up. A careless end-to-end movement of the frame will crush bees between the frame ends and the chamber sidewall. Set the first frame outside the chamber to preserve the working space inside the chamber while examining the remaining frames.

When you have completed the hive exam, remember to replace the frames in the same positions where they were positioned when you started the examination. Shuffling frame positions disrupts the normal brood nest organization; it's like putting the sofa in the kitchen. After the inspection, when the frames are back in their original positions, use your hive tool to squeeze the nine combs back into the middle of the chamber to preserve the exaggerated bee space on both sides.

In time, the accumulation of propolis at the frame contact points will space the frames perfectly. Permanent frame spacers are not necessary, and they actually interfere with the hive-inspection procedures by preventing the frames from sliding side-to-side and allowing you to manipulate the workspace between frames.

Variations in chamber size due to different manufacturers may prevent you from using the aforementioned procedures. (Unfortunately, bee-equipment manufacturers have not agreed to standard dimensions for hives.) In some cases, you may have to use ten frames in a chamber. Otherwise, the spaces between the chamber walls and the frames may allow the bees to fill the space with distorted, unusable combs.

## Magic Smoke

You'd think, considering our modern technology, that there would be some way to treat bees to put them in a warm and fuzzy mood or to harmlessly anesthetize them for a few minutes during hive manipulations. But that's not the case. The invention of the bee smoker in 1875 was a landmark advance in beekeeping tools. It's not high-tech, but it works fine if you know how to use it properly. Understanding how to generate and use smoke is the difference between enjoying your bees and constantly engaging in combat with them. Use quality smoke properly, and you are in control.

Smoke doesn't calm the bees any more than tear gas calms an unruly crowd, but it does confuse them. Bees' olfactory senses are incredibly sensitive to odors. The sudden explosion of noxious smoke in their environment creates immediate havoc throughout the colony. Reactions to the smoke are instantaneous, almost completely overriding defensive behavior for the moment. Disorganization is rampant, and you can hear intense fanning and buzzing. (Listen carefully, and you might just hear thousands of little sneezes, coughs, and bee profanities!)

Use a good fuel that contains natural oils and resins that are pleasant, not acrid. Beekeepers use a great variety of smoker fuels, such as rolled-up burlap bags, rolled corrugated paper, compressed wood-chip pellets, compressed cotton, and even dried animal dung. Most of these don't produce good-quality smoke, and some may contain harmful chemicals. A very convenient and economical smoker fuel is pine shavings, normally used as bedding for animals and available at animal-supply stores. Make sure there are no chemical additives, such as a fire retardant. Do not compress the chips inside the smoker; the fuel needs to breathe.

Pine needles are excellent fuel but are physically difficult to stuff into the smoker, especially long-leaf varieties. If you have access to pine needles, you can pile them up and run over them a few times with your lawnmower to chop them into small pieces. Pine needles can be used alone or mixed with the pine shavings. Pine needles produce excellent-quality smoke, but when used alone, they tend to burn too fast and create excessive amounts of creosote (chimney tar) that eventually gums up your smoker. When this happens, you can clean the smoker by continuously puffing it until all the fuel has been burned and the smoker is nearly red hot. Tar and creosote lining the barrel will turn into a brittle carbon crust that you can easily remove.

## SIZE MATTERS

Proper use of smoke is your first and most important means of sting prevention. By pumping the bellows of the bee smoker, large quantities of smoke can be delivered on target and quickly. Smokers are available in several sizes. Hobby beekeepers frequently buy small smokers, which is a big mistake. Large smokers burn much longer and produce a better quality smoke.

## How to Use the Bee Smoker

Lighting the smoker is a daunting task for some people. It can be almost comical to watch a beginner try to ignite the fire on top of the fuel. Here's an easy method: place 1 cup of dry wood chips in the bottom of the open smoker, then drop a lighted match or small piece of burning paper on top of the fuel. Allow at least half of the fuel to burn and produce a bed of hot embers, and then fill the chamber to the top while puffing the bellows. Close the lid and continue puffing to produce great clouds of *cool, dense, white smoke*. A large smoker generates cool, dense, white smoke—worth repeating— qualities that cause maximum response by bees.

The secret to preventing stings lies in the quality of smoke and how it's used. Cool, dense, white smoke works beautifully.

Don't forget to keep your smoker filled as you work. Otherwise, it will gradually start burning too hot and produce poor-quality smoke, which will be hot, bluish, and less dense. Hot smoke may smell acrid and may even aggravate the bees. How would you like to have your delicate antennae singed if you were an innocent little bee? Plus, it's just not as effective as cool, dense, white smoke. Once you learn to produce and properly apply quality smoke, you can take command, be the master of the bees, and impress fellow beekeepers with your prowess.

## Opening the Hive Safely

When you open a hive, always stand to the side. Standing in front of the hive entrance is comparable to standing at the end of the runway at an airport while planes come in. Returning bees orient to the image of the hive. Your presence changes this image and confuses them. They need to see their hive entrance so that they can enter immediately, not accumulate in the air around you. With the finesse of a snake charmer, position yourself at the side of the hive that is *downwind* from the hive. (*Hint*: downwind is the side where a puff of smoke blows away from the hive.) Place the smoker nozzle about 2 inches from the entrance so that the smoke will go *inside* the

When you need to critically examine cells—for example, when looking for eggs—you must hold the comb so that the sun is behind your back, illuminating the interiors of the cells.

hive. Puff hard repeatedly while you sweep the nozzle from one side of the entrance to the other and back again. Smoke must be forcefully blown inside the hive just before you open it. Simply blowing smoke *at* the hive entrance from a distance isn't effective.

*Here is a big secret for preventing stings*: wait approximately three minutes after puffing the smoke inside the hive entrance before opening the hive. Most bees react to the smoke by ducking into open cells and engorging on nectar and honey. No one knows why—it's just instinctive. Bees filled with honey are more docile, which is a good thing for you. Once engorged, they are far less likely to sting. While you are waiting, smoke all the hives in the apiary. You'll be surprised and pleased to see how well this strategy works.

Just before you remove the hive cover, smoke the entrance again. Smoke under the cover as you are lifting it and as bees are being exposed. Remove the cover slowly in one motion (be careful to prevent the cover from dropping back to its initial position and crushing bees). Smoke bees inside the hive as needed to keep the frame tops relatively clear of bees as you proceed to remove the combs.

Now that the bees are confused and under your control, you are the master. Be alert; they will soon recover, sometimes rapidly. You have to be the judge as to when you need that next puff of smoke to maintain control. When they begin to return to their original positions on top of the comb frames—watching you watch them—they need another puff. However, be careful not to overdo it. Don't use more smoke than is necessary to keep control.

It bears repeating that you should not rely on a bee suit or protective clothing as your primary defense against stings; proper use of smoke is your first line of defense. Protective clothing serves as your backup protection. Otherwise, you'll feel invincible in your suit of armor, develop false security, take chances, and never develop finesse in colony-manipulation techniques. Under these circumstances, you'll soon discover a cloud of defensive bees flying around your head. You will be a threat to yourself as well as to your neighbors, whose kids and pets may be playing in a nearby backyard.

**This is what you see when you first open your hive. Smoke the bees and they'll go down onto the combs. This smoke is very poor quality—definitely not the cool, dense, white smoke you should use.**

## Bee-Smoker Hazards

Many beekeepers have lost their vehicles, buildings, and even their homes because of their careless use of bee smokers. Fire prevention is extremely important, especially in arid areas where wildfires are a threat. If a spark filter isn't included with your new smoker, and you use your smoker in a high-fire-risk environment, you should make a spark filter. Cut a circular filter from #8-mesh galvanized hardware cloth in a size that can be wedged just inside the hinged lid of the smoker. This simple filter prevents sparks and pieces of fuel from blowing out. In areas where there is a fire danger, always have a water sprayer handy. It should be pressurized and tested before you light the smoker so that you can use it instantly. When your work is done, you can quickly extinguish the smoker by plugging or taping both openings—the air-intake hole at the bottom and the exhaust hole in the nozzle at the top—with tape or other nonflammable material. The accumulation of carbon dioxide quickly extinguishes the

**This simple #8-mesh-wire screen filter prevents the blowout of fuel particles and sparks that could start a fire.**

fire. A further precaution is to place the smoker inside of a metal container with a tight lid.

When you're positive that the fire is totally extinguished, dump the ashes into a metal container. Do not use water on the ashes. Water and ashes would react chemically to produce some very corrosive chemicals. Opening a hive for routine inspection would be much easier if you had three hands: two for grasping each comb frame and the third for holding the smoker. Beginner beekeepers often awkwardly place their smokers on the ground, hang them on the sides of the hives, place them on nearby hive tops, and so on. Remember that the smoker is your best defensive weapon—you need it fast, you need it handy, and you don't want to make excessive motions that could stimulate bees to sting. Here is a tip: learn how to hold the smoker by pressing the bellows (and only the bellows) between your knees as you work. Practice this away from the hive and before you light the smoker; otherwise, you may get a painful burn.

## Africanized Bees

A discussion of honey bee sting risks would not be complete without comparing the defensive behavior of Africanized honey bees (AHBs) with the common and ubiquitous European honey bees (EHBs) kept by beekeepers in the United States. African bees were introduced experimentally in 1957 into Brazil, where they mated with European bees, producing bees that are very hardy and extremely defensive. The resulting hybrids, called Africanized bees, gradually migrated northward to the United States and are now established permanently in the southern states. They require warmer weather than European honey bees do. During their evolution in Africa, they were subjected to intensive predation. Consequently, they developed the incredibly defensive behavior that now defines Africanized bees.

AHBs are extremely sensitive to disturbances at or near their hives, far more so than EHBs are. When a typical EHB colony is disturbed, very few bees participate in defensive behavior, and those that do typically defend an area around approximately 10–20 feet from the hive. A similar disturbance to an AHB colony causes thousands of defensive bees to instantly emerge from the hive and defend a very large area—up to a quarter-mile in all directions from their hive. Within this area, they will sting and potentially kill animals and people. The potency of an AHB stinger is identical to an EHB stinger, but AHBs are a greater risk because they deposit hundreds to thousands of stings, thereby delivering massive amounts of venom. Other than in defensiveness, AHBs are identical in appearance to EHBs. They produce honey, and they pollinate well.

Hobby beekeepers should take every precaution to avoid the introduction of AHB stock into their hives. Don't start new colonies with swarms caught in areas infested with AHBs. A high percentage of these swarms will be Africanized, headed by queens that have mated with Africanized drones. Colonies established with an AHB swarm typically will not be very defensive for the first few weeks, but one day—during normal hive manipulations that previously had been pleasant—the population will suddenly explode in a defensive frenzy that is extremely dangerous and nearly uncontrollable. Hobby beekeepers in AHB areas should purchase mated queens from professional queen-bee breeders located in areas free of AHBs. There must be no doubt concerning their pedigree. Furthermore, these queens should be identified with an ID tag or a paint mark on the thorax or by clipping a wing. If your hive is located where AHBs are established, and there is evidence that an established

queen is missing or has been replaced by a new queen of unknown pedigree, then a new replacement queen must be introduced as soon as possible.

AHBs are not fun and can be very dangerous for the hobby beekeeper. Liability exposure is also a major consideration when AHBs are involved. Stinging incidents seriously prejudice people against bees and undermine legitimate beekeeping operations everywhere.

**This bee could be an Africanized bee or a European bee. Identification based on appearance alone is not possible.**

# CHAPTER 9

# How to Manage Colonies

Colony management is the most controversial topic in the world of beekeeping. Have you ever tried to tell other parents how to raise their kids? That would be a simple task compared to trying to tell beekeepers how to manage their colonies throughout the year. A single recipe for annual management is not possible because the system contains too many uncontrollable variables— climate, nectar and pollen plants, genetics, objectives, knowledge of bee behavior, and so on. Despite this scenario, you should observe some basic management practices in almost all areas. There is only one really good way—your way—to manage your bees to fit your situation, and this will change as you acquire experience. It's a good idea to keep records so you won't forget next year what you learned this year. Events, such as the onset and duration of honey flows and the appearance of queen cells, tend to repeat each year. You can make better preparations if you know when to expect these events.

# Potential Problems with Neighbors

There is one bee-related problem that is real but not serious, except in certain situations. Bees normally defecate while flying. Feces, composed primarily of indigestible pollen "shells," are voided in one sudden burst. The most obvious deposits are on the shiny paint finishes of vehicles. If inclement weather delays bees' regular flights, feces accumulate in their guts, leading to a shower of fecal droppings when favorable weather arrives. The frequency of droppings decreases as the distance from the hive increases. Although the droppings do not normally damage the paint surface on cars, they require more effort to remove than, for instance, dust. After a brief soaking, the little brown spots can usually be removed easily.

Be considerate of your neighbors by locating your hives as far away from them as possible. A complimentary jar of honey now and then, *ideally before any problems arise*, is usually a very effective peace offering. In rare circumstances, a neighbor may have a known history of anaphylactic reaction to insect stings or an irrational fear of insects. It is to your advantage to respect these circumstances. Be creative and keep your hives down the street, in the backyard of a friend. You should make every effort to avoid litigation—or even the hint of litigation—at all costs. A sting now and then is okay for you, the bee lover. But that same sting on a neighbor can create an avalanche of problems that can end your hobby in its infancy.

Despite your best efforts, some events are beyond your control. What if your neighbor is stung by a yellowjacket on his property, but he doesn't know the difference between

**Nice neighbors can become even nicer if you surprise them with a bottle of honey now and then.**

yellow jackets and honey bees and he honestly believes that your bees are the real culprits?

Overall, the odds of keeping your bees successfully and without incident are very much in your favor if you are aware of the pitfalls and take sensible precautions. Isolating your hives from view can be helpful under some circumstances. Installing a high, solid fence around your hives may be practical, not only from the standpoint of blocking the view but also for directing bee flight upward and away from your yard. Attempts to keep your beekeeping activities secret rarely work.

**Here's a typical backyard hobby bee apiary with a fence separating yards. Bee flight is directed up and away from neighbors.**

This is especially true if swarms from your hives cluster on your neighbor's property.

Some beekeepers abuse the privilege of keeping bees in urban environments by keeping too many hives. If you live in such an area, you should keep a maximum of two hives unless your lot is unusually large.

## Arrangement of Hives in Apiary

The spatial configuration of hives in your apiary is extremely important. In nature, bee colonies are sparse, scattered over large areas. This distribution minimizes a host of problems, such as competition for nectar and pollen as well as interactions between colonies that could lead to the spread of diseases. Do not crowd hives too closely together in your apiary.

In the hobby-beekeeping scenario, assuming that space is available, there should be a minimum of 5 feet between colonies, and more is better. Why? The main reason is to provide an adequate work area for hive manipulations. Removing frames of comb during inspections requires the beekeeper to stand at one side of the hive, never working from the front or back of the hive. Another reason for adequate spacing is to preserve the option of standing downwind from the colony during manipulations. You'll also require workspace whenever you need several stacks of hive equipment for certain operations, such as harvesting honey. Spacing hives apart also minimizes undesirable colony interactions. For example, if hives are placed too close together, some bees returning after flights frequently enter the wrong hive. Beekeepers call this *drifting*. Some of the drifters transfer to the new hive permanently. Bee transfers between hives increase the chances of spreading diseases and parasites. Maintain proper work space between hives.

## Producing High-Quality Combs

Frames of quality comb are the backbone of bee colonies. They provide food storage, a space for brood rearing, and a vertical floor where most activities occur. When producing new combs, you should place frames of foundation in colonies during late spring and summer. Optimum conditions for producing new combs are: (a) a very populous colony crowded with bees, (b) warm weather to support active foraging, (c) placement of the super containing new frames with foundation immediately above the brood chamber, and (d) an active honey flow. If natural nectar is scarce, you can feed sugar syrup to stimulate comb construction.

When producing new combs, place ten frames, squeezed together tightly, in the super, with the remaining space equally distributed between the outer frames and the chamber walls. Sometimes bees are slow to move into the new equipment to start comb building. You can accelerate the process by placing an older comb in the center of the super to attract bees and stimulate the onset of comb building. When cell construction and honey storage are finished, you can harvest the honey super. Be cautious during the extraction process, because new combs are fragile.

If you are using beeswax instead of plastic foundation, you can enhance the strength of new combs by temporarily placing them in a honey super located immediately above the brood chamber(s) and just beneath the queen excluder. Leave the combs in this position long enough for the rearing of several cycles of brood— perhaps for two months. Then place the queen excluder between the super and the brood chambers and put the queen beneath the queen excluder. As brood emerges, the bees will fill the cells with nectar and honey. Combs are strengthened by the accumulation of cocoons on the interior cell walls. Now you have durable comb that is light in color—ideal for use in honey supers. When the old brood comb needs to be replaced because of age, simply replace it with the honeycomb.

# Seasonal Cycles in Colonies

Let's examine some of the basics that are involved in developing a one-size-nearly-fits-all management program for your hobby operation. Variable climates and floral food sources greatly affect management strategies in different geographic areas. Hives in upstate New York may be buried in snow at the same time that colonies in Florida are actively foraging and making honey. Management procedures are similar in both locations, but the timing is different.

Virtually all of the seasonal activities in a bee colony are repeated each year. Plants that produce nectar and pollen also have similar seasonal cycles. Bee activities and plant growth are in harmony because they are controlled simultaneously by changes in day length and weather conditions. Short-term changes in available nectar and pollen have profound effects on the development and health of bee colonies as well as on the production of honey. Beekeepers must learn about these seasonal changes and their effects on colonies in order to develop management strategies that maximize honey production. What to do—and when to do it—defines your success as a beekeeper.

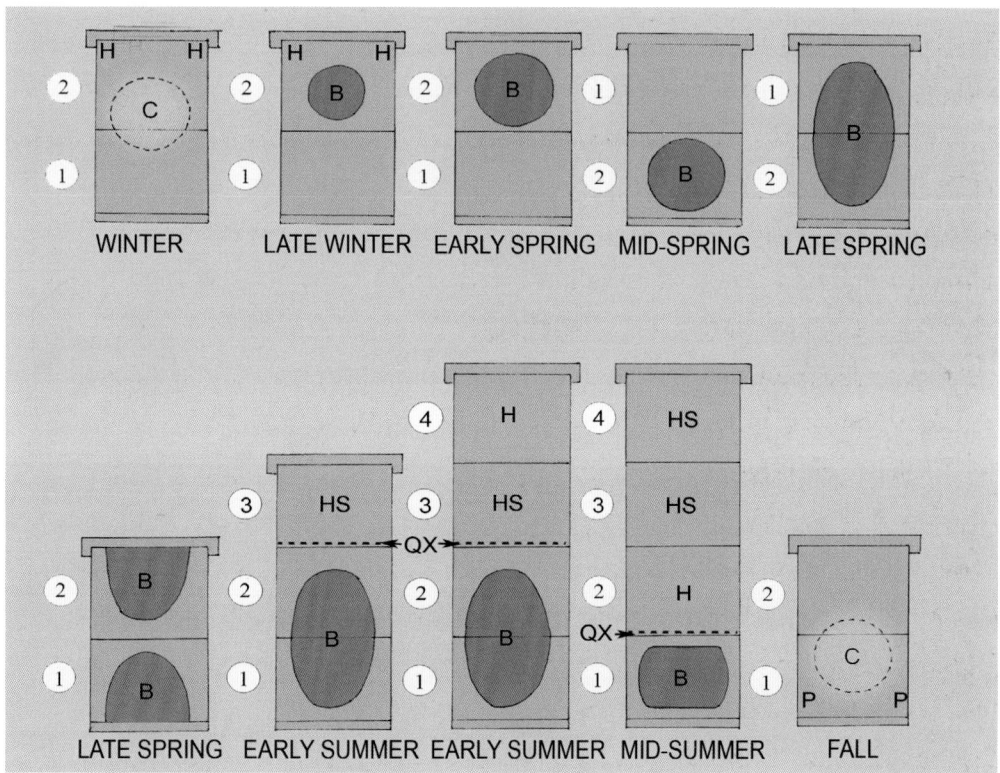

Planning ahead greatly simplifies beekeeping. With minor modifications, these seasonal management strategies are effective, especially in colder climates where brood rearing ceases in the winter. (B: brood; C: broodless cluster; QX: queen excluder, HS: harvestable honey in honey super; H: honey for overwintering food. The numbers in yellow identify chambers as their positions change.)

# Basic Seasonal Management

An annual seasonal management summary is helpful for visualizing the configurations of hive chambers throughout a typical year (see the graphic on the previous page). The management summary that follows has to be adjusted to correspond with climatic conditions in your area.

## Winter

Bees do not hibernate. Any time that the temperature is above approximately 55 degrees Fahrenheit, there can be active flight. In the southern states, bees are active throughout the year. In the northern states, they may be confined inside their hives for several months with no flight. Bees have an incredible ability to store feces in their rectum. Normally, they defecate while flying, so when that warm day finally arrives, the bees will venture forth from the hives for brief "cleansing flights." Loads of feces are dropped like tiny bombs in midair—sometimes discoloring the snow in nearby areas. One can imagine hearing the little sighs of relief. A little advice: this is not a good time to be near the hive.

In most areas, bees overwinter in hives composed of two deep chambers, where they keep warm by clustering tightly. They survive by eating honey and pollen stored during the previous summer and fall months. As they consume the honey, the cluster gradually moves upward onto new honey stores. Inside the broodless cluster, the temperature is 43 to 46 degrees Fahrenheit. Bees exchange positions within the cluster to stay warm; those on the outside of the cluster burrow into the center to get warm and are replaced by bees from the inside.

Longer days trigger brood-rearing activity. The queen's abdomen enlarges once more as her huge ovaries fill with eggs. Workers metabolize honey at a much higher rate and elevate the internal cluster temperature to 93 to 94 Fahrenheit. Their tightly packed bodies function as a blanket to minimize heat loss from the core of the cluster, where brood is developing. The process of replacing the old overwintering bees has begun. By late winter or early spring, the cluster is usually located primarily in the upper chamber.

## Spring

When the upper chamber becomes fully occupied by bees and brood, usually by early spring, reverse the two chambers so that the upper chamber with brood becomes the lower chamber, resting on the bottom board. Reversing the chambers allows rapid upward expansion of the brood nest while maintaining brood-rearing in the lower chamber. As spring progresses, colony population growth accelerates rapidly in response to increasing availability of nectar and pollen. Spring flowers such as dandelions provide a wonderful food stimulus. The result is a population explosion. Now the colony is packed with bees, filling the entire available space. This is the time of year when swarming naturally occurs (see Swarm Prevention in this chapter).

By mid-spring, a second reversal may be appropriate, depending upon the brood quantity and the population of the developing colony. During late spring, when both chambers are teeming with bees and there is abundant nectar foraging, you should add a honey super on top, expanding the hive to three chambers. Place a queen excluder between the super and the two brood chambers to prevent further upward expansion of the brood nest into the honey super.

## Summer

Most of the harvestable honey is made during the summer. Nectar-foraging activity climaxes because days are longer, weather is more favorable, many nectar-producing plants are blooming, and colonies have developed large populations of foragers. This is the season that hobby beekeepers love—stacking supers on their colonies and watching them fill up with honey. Sometimes there may not be a honey flow. It's always a gamble.

Throughout the remainder of the active season, you should add honey supers for storage of incoming nectar and stored honey whenever the uppermost honey super becomes about two-thirds full of nectar and honey. One option is to insert the empty honey supers immediately above the queen excluder, under the super that is full of honey. This involves more work, but some beekeepers claim that "bottom supering" increases honey production.

**High hives are risky and difficult to manage. Harvest honey more frequently to avoid skyscraper hives.**

Do not add supers unless there is a honey flow in progress or one is anticipated in the near future. Excessive honey supers burden the colony by necessitating additional maintenance and perhaps reducing the efficiency of regulating temperature and humidity inside the hive. In most circumstances, your hive should have a maximum of two deep honey supers, or three to four shallow supers, at any given time. Deep supers full of honey are very heavy—around 80 pounds—and they're difficult to lift when the hive is too high. (Remember, you never have to lift a full super of honey—just remove individual frames of honey for ease of handling.) It's best to harvest honey periodically to prevent the hive from getting too high. Extract the honey and immediately return the empty combs to the hive for additional honey storage, provided that the honey flow is still in progress.

Regional differences in the time and intensity of honey flows will dictate the optimal time to harvest honey. When harvesting in late summer or early fall, be sure that the upper chamber has enough honey and pollen for winter stores. Here is one way to ensure adequate winter food stores: around mid- to late summer, when brood-rearing naturally starts to decline, confine the queen to the lower brood chamber for the remainder of the honey-producing season. You don't have to find her. Just shake and brush the bees and queen into the lower chamber and confine her there with an excluder. One chamber is enough space for brood-rearing at this time of year, provided that all combs are accessible for brood-rearing.

**Wheels are handy for transporting heavy honey supers.**

## CHAMBER REVERSAL

You have the option to reverse the brood chamber positions at any time during the active brood-rearing season whenever you notice significantly more crowded conditions in the upper brood chamber compared to the lower chamber. The tendency to expand brood-rearing upward frequently leads to an underutilized—sometimes almost abandoned—lower brood chamber. Position reversal promotes better utilization of both brood chambers.

When the queen is confined to one chamber, exchange frames of honey in the lower chamber for frames of brood removed from the upper chamber so that all combs in the lower brood chamber are accessible for brood-rearing, not blocked by stored honey. As brood matures and adult bees emerge in the second brood chamber above the queen excluder, the resulting empty cells will provide abundant space to store honey for winter food. Pollen gathered in early fall will be stored in the lower two brood chambers, where it will be available during the winter. At the appropriate time, you can harvest all of the honey stored in the honey supers above the second brood chamber without the hassle of finding the queen and sorting combs of honey and brood.

## Fall

Some nectar and pollen plants bloom during early fall. Beekeepers treasure a fall honey flow because this is the last opportunity for colonies to store honey for winter food. All honey supers must be extracted before cold weather arrives. After honey has been harvested for the season and the hive has been reduced to the two brood chambers, be sure to remove the queen excluder. Otherwise, as the cluster moves upward during the winter months, the queen may be trapped beneath the queen excluder where she may perish.

In the early fall, brood-rearing activities slow dramatically and may stop completely by November in cold climates, or they continue at a reduced level in warm climates. The survival of overwintering colonies is contingent upon having (a) a large population of healthy bees, (b) 40 to 60 pounds of honey, and (c) a healthy queen. By late fall, beekeepers in cold climates normally reduce the hive size to two standard deep chambers—the amount of space that bees can properly care for during the winter.

# Monitoring Hive Conditions

Periodic examinations are necessary to monitor colony conditions, especially during the spring, when the colony's population changes dramatically as the bees prepare to swarm. Hives should not be disturbed during the winter months except in warmer climates. Delay the post-winter examination until bees are actively flying again.

It's always exciting to see how your colony is doing, especially if everything is going well. Dog owners pet their dogs and beekeepers open their hives—same concept. It's fun. Enjoy the experience, but don't just gaze admiringly at the beautiful sight of bees working together. You have to do more than just smile and congratulate yourself for choosing the best hobby in the world, along with dipping your spoon into the comb for a taste of fresh honey.

While the hive is open, look for specific indicators of colony health, such as adequate food stores, population levels, and brood-rearing status. And—just for fun—see if you can find the queen. Don't feel concerned if you don't see her at each visit. Sometimes even the experts have trouble finding her highness because she blends in very well with the worker-bee population. Sometimes she runs and hides—or so it seems—in unlikely places on the interior walls of the brood chamber. Having the queen marked with a visible identification tag or paint mark is helpful, but you don't have to see the queen to evaluate her performance. Just examine the brood combs to determine if there is enough brood in various stages of development to confirm the queen's presence and vitality. If you see eggs deposited normally, one per cell, you know that she is present (or at least has been within the last three days). Critically examine the brood pattern, especially on frames where the brood cells are capped. There should be very few uncapped cells in the capped brood area.

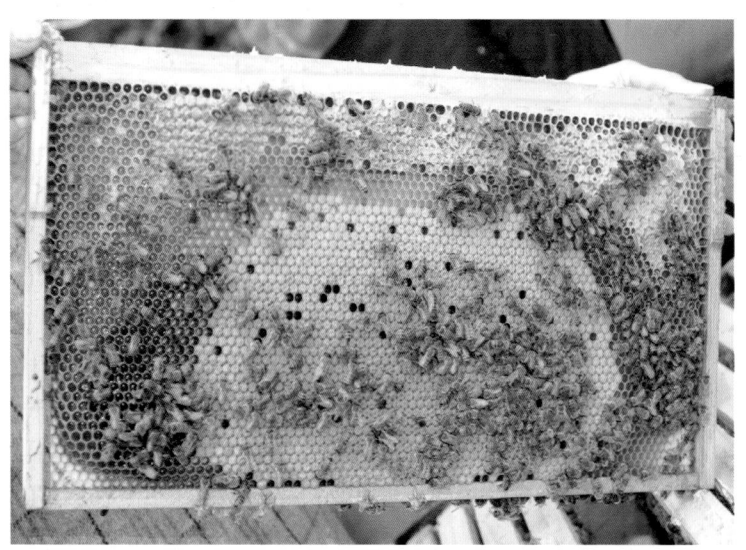

A compact pattern of capped brood cells with very few open cells is evidence that the queen is in good health and reproductively active. The result is a vigorous, productive, populous colony.

## Protecting Bees from Themselves

Hives should not be open any longer than is necessary during inspections. Foraging bees are opportunistic. Scout bees are constantly searching for new sources of food, and exposed honey is the most stimulating food possible for a honey bee. During hive manipulations, you expose the honey stores. Scout bees from other colonies may discover the free meal and quickly recruit additional foragers. Within minutes, large numbers of foragers will start collecting nectar and honey from the exposed combs. Beekeepers call these *robber bees*.

Returning resident bees alight and enter their familiar hive entrance without hesitation. However, robber bees are not familiar with the entrance, and they search for honey by smelling it. Consequently, you'll see them hovering near the entrance and flying around the sides and back of the hive, persistently trying to gain entry anywhere there is honey odor leaking from the hive, especially at the cracks and joints between chambers. During intensive robbing, you can observe the defensive bees at the hive entrance fighting and stinging the invading bees. Populous (strong) colonies may be able to defend themselves, but weak colonies can be overwhelmed by the invasion.

The threat of robbing is minimal during a honey flow, when most of the foragers are already dedicated to known food resources away from the hive. When nectar is not available, the potential for serious robbing greatly increases. If you own just one hive, it's unlikely that scouts from distant colonies will discover your hive during the brief time it's open. Multiple hives within your apiary are the problem. Once foragers from neighboring hives learn to rob, their behavior can become a chronic problem. Each time hives in the apiary are opened, robber bees pounce on the vulnerable colonies, and robbing activity just gets worse. It's an especially important consideration for those beekeepers who like to open their hives very frequently.

You can eliminate or at least minimize robbing behavior by using several strategies. As the saying goes, an ounce of prevention is worth a pound of cure. First, avoid hive manipulations as much as possible during a nectar dearth, when foraging activity is meager. Second, cover exposed chambers with screen covers on the bottom and top so that they become decoy odor sources that will keep the potential robbers busy but without food reward. Third, be alert to robbing behavior as you are inspecting the hive so that you can close the hive quickly, before the problem gets serious. Fourth, reduce the hive entrance opening to approximately 2 inches by blocking the remaining part of the entrance with #8-mesh galvanized hardware cloth (screen). This tactic is even more effective if you install the mesh before robbing begins. Again, odors pass through the mesh and misdirect the robber bees away from the functional entrance. Guard bees will concentrate at the small opening and be able to defend the functional entrance more efficiently. The mesh will permit adequate ventilation, and the resident bees quickly learn the location of the reduced entrance. Leave the mesh in position for the duration of the nectar dearth, when robbing is most likely. Do not use a solid barrier, such as a piece of wood, to reduce the entrance. This aggravates the robbing problem because the odors coming exclusively from the small entrance would actually direct the robber bees to the entrance. Solid barriers also restrict ventilation.

**A ventilation screen has many uses, such as providing ventilation during hive transport and preventing robbing when chambers of exposed combs are covered.**

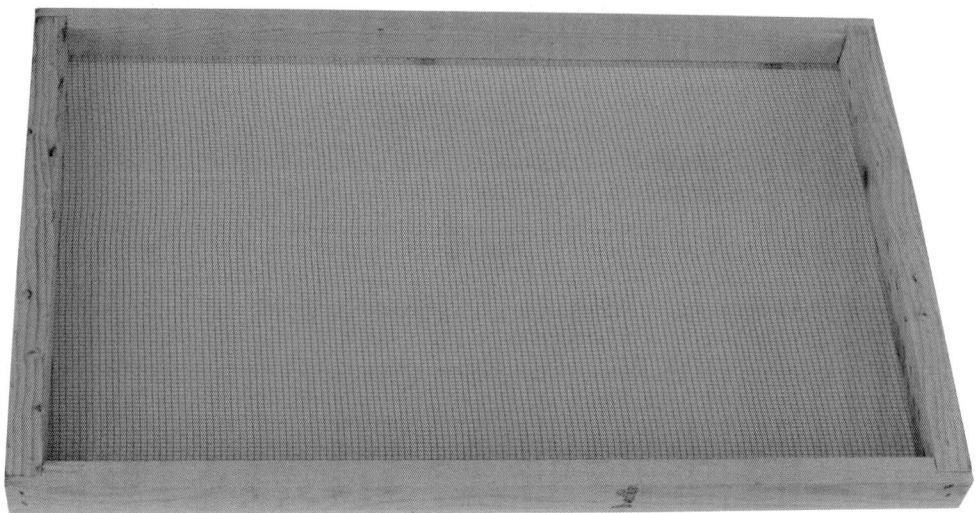

## Brood Pattern

When the queen is laying eggs, does she just move around randomly and a lay one here and one there? No; she tends to start near the middle of a comb and expand in all directions where there are empty cells. The age and distribution of brood on both sides of the comb is almost identical—a mirror image. If the queen is performing normally, she lays one egg in each cell. She will miss a cell now and then because it may be occupied by pollen. As the brood develops, some larvae may die for various reasons, leaving open cells in areas where most cells should be capped. When developing brood is old enough to be capped—about nine days after the egg is laid—the percentage and distribution of open cells in the capped brood area may indicate whether or not there is a problem that needs attention. Maybe some of the eggs were not viable because the queen bee is failing and needs to be replaced. Maybe disease-causing microorganisms are killing the larvae or pupae and the bodies have been removed by housecleaner bees. Attempting to diagnose this kind of problem when you first begin beekeeping is very risky—akin to humans who try to self-diagnose their illnesses. The novice beekeeper should share diagnostic opinions with an experienced beekeeper. You

**Scattered cappings in a brood comb should sound the alarm: "Something is wrong! Diagnose immediately."**

must address the problems in a timely manner because, without a normal rate of reproduction, the population of worker bees could decline until the colony dies.

## Food Stores

At least 10 pounds of reserve honey—the equivalent of about two deep frames of honey—should be in the hive at all times. Maintaining minimum honey reserves during the active foraging season is not usually a problem for the hobby beekeeper. The greatest risk of starvation is during intensive brood-rearing in early spring before nectar sources are available. A greedy beekeeper who harvests too much honey at one time can create another potential starvation risk. Harvesting all of the honey at once is especially risky when an anticipated post-harvest honey flow doesn't materialize, owing to bad weather, drought, or other factors. You can rescue a starving colony by feeding sugar syrup.

Pollen reserves are more difficult to evaluate. You should be able to see dozens of pollen-containing cells in the brood chamber. Pollen deficiency is a greater risk when large apiaries—perhaps fifty hives or more under commercial conditions—are seriously competing for limited pollen sources. Hobby beekeepers with a small number of hives (perhaps less than five) usually don't need to be concerned because the pollen sources within flight range are divided among so few colonies that they will almost always have adequate pollen reserves. This is especially true in urban environments, where there is a greater diversity of pollen sources and less competition from large apiaries in the area. Manufactured pollen substitutes are available—and fun to use experimentally—but rarely would be necessary for the hobby beekeeper.

## Worker Bee Population

Experience will teach you how to recognize normal bee populations as they fluctuate during the year. Two pounds of bees (approximately 8,000 bees) in your first hive may seem like a lot to you, but just wait until you have 40,000 bees in the hive several months later. How can you be sure that your hives have normal populations? It is helpful to have multiple hives in the same apiary to compare with each other. If you've managed the hives identically, and one hive is twice as populous as the other, then you can look for problems in the weak colony. The most frequent cause of a weak colony is a poorly performing queen. Maybe she is old and has lost her vigor. Simply replace the queen in the weak colony and see what happens (see Queen Replacement and Introduction in this chapter). Maybe you'll get a much better queen and restore the colony population.

**Large populations greatly enhance the performance of colonies. Beekeepers would say this image depicts a strong colony because of the obvious high population density.**

## Adequate Comb Space

Good management for honey production requires adequate hive space to accommodate bee populations as well as the storage of surplus honey in amounts far beyond the bees' nutritional needs. Sometimes hives are as high as your head. A hive may contain 100 pounds of honey—or even more—at one time. Large crops of honey cannot be produced unless you provide enough empty combs at the right time.

# Feeding Sugar Syrup

Colonies can be fed ordinary table sugar (sucrose). This is an excellent carbohydrate source that substitutes perfectly for honey. It is a highly refined natural product that is pure, free of microorganisms, economical, normally found in nectars, and totally digestible by honey bees. One reason for feeding sugar is to compensate for the loss of reserve honey in the hive. As previously mentioned, there should always be at least 10 pounds—two standard frames—of honey in the hive. Honey stores may be depleted when honey flows don't materialize or when beekeepers harvest more than they should, gambling that the next honey flow will replenish honey stores.

Feeding is also a means of stimulating brood production at critical times when nectar is not available. Early spring stimulatory feeding accelerates brood-rearing and helps to produce populous colonies earlier in the season. The bees may need early fall feeding to provide enough food reserves to survive the winter.

You can mix sugar syrup in various concentrations, depending upon what it will be used for. If the purpose is to stimulate colony growth in established colonies or comb construction in new colonies, then feeding dilute sugar syrup is appropriate. Use 1 part sugar to 2 parts water (approximately 1.5 pounds of sugar per gallon of water). For supplementary feeding in the fall, a concentrated solution is better. Use 2 parts sugar to 1 part water (approximately 5 pounds of sugar per gallon of water). You don't need to add flavors—bees are little gluttons and will readily take the plain syrup. Do not add honey to the syrup, because its attractive fragrance may stimulate robbing behavior. Honey also contains microorganisms that encourage spoilage if the syrup isn't taken quickly. If in doubt, just feed syrup made of equal parts of sugar and water (about 3 pounds of sugar per gallon of water), and you will not hear any bee complaints.

You can find various kinds of bee feeders commercially. The *division-board feeder* has approximately the same dimensions as a frame of comb and actually substitutes (during the feeding process) for a brood-chamber frame in the outside position next to the hive wall. Functionally, it is a small tank that is open at the top for bee access and

**A division board feeder, open at the top to permit bees access, is filled with sugar syrup.**

contains up to a gallon of syrup. Division-board feeders have several minor disadvantages. You have the inconvenience of having to open the hive to insert the feeder and to monitor the syrup status. Additionally, a few bees are likely to end up in the syrup.

A *gravity feeder* is a popular alternative to the division-board feeder. You place syrup in a container—usually a quart-size Mason jar or large plastic container—that has a lid perforated with small holes. You then invert the container and place it where bees can suck the syrup through the tiny holes with their proboscises. After the first few drops of leakage, a vacuum forms. As the bees feed, tiny bubbles relieve the

Feeding through the hive cover works well and uses the least equipment. The feeder is visible to indicate the time to refill. If there is leakage, the syrup drips inside the hive, where it is immediately consumed.

If the hive has a telescoping cover, you can feed from the top. The container pictured is ajar to show that the feeder is placed directly over the inner cover hole.

Partially screening the entrance instead of blocking it solidly misdirects potential robbers to hive odors in the screened area and reduces the area to be protected by guard bees.

vacuum slightly, just enough to allow the release of a few drops.

If you're using a wooden hive cover, a gravity feeder can be placed on top of the hive in a hole made to accommodate the diameter of the feeder lid. When using a telescoping hive top, you can enclose the feeder in an empty chamber resting on an inner cover that has a hole to provide bee access to the lid. This is the easiest way to feed sugar syrup. Bees take syrup faster when it is presented on top of the hive.

Some hobby beekeepers use an *entrance gravity feeder*. This gizmo accommodates the perforated lid of the feeder and inserts snugly into the entrance opening. (Don't forget to smoke the entrance before you insert the feeder.) Monitoring the syrup uptake is easy because the container is always visible outside the hive. In most situations, feeding from the hive top is better because entrance feeders tend to leak, sometimes attracting ants or robber bees. Do not use the entrance gravity feeder in cold weather, when there is no bee flight at the entrance.

## Feeding Pollen Substitutes

There is a long history of attempts to formulate pollen substitutes that contain the complex nutrients found in pollen. Those who have tried have had limited success, as indicated by various commercially available pollen substitutes.

Beekeepers have experimented with every imaginable pollen-substitute recipe. One do-it-yourself option is to install pollen traps on hive entrances to collect fresh pollen when it is plentiful—especially during the spring—and store it in sealed bags in the freezer. Later, you can knead it like bread dough into a fondant candy made with one part confectioner's sugar and one part honey. The concentration of pollen is not critical—10 to 20 percent is fine. If you squish this soft candy onto and between the top bars of the brood chamber, the bees will quickly take it.

Commercially prepared pollen substitutes are available in bee-supply catalogs and on the Internet, but they are not as nutritious as natural pollen. If you enjoy feeding animals and get pleasure in "mothering your little hungry ones," then why not experiment? That's the joy of hobby beekeeping. Fortunately, hobby beekeepers don't need to be too concerned about feeding pollen substitutes because there usually is enough pollen in the immediate area to support the needs of small numbers of colonies.

**Fall flowers help bees store enough honey for overwintering.**

# Sacrificing Bees Humanely

There are situations when bees need to be sacrificed for the safety of humans and animals. For example, beekeepers should sacrifice as quickly as possible Africanized bee colonies near human and animal habitations. Africanized bees are very dangerous. Another situation that justifies extermination is the development of laying workers that cause the colony population to decline to the point that the bees cannot protect the combs from destruction by insect pests. Because a laying worker colony's demise is imminent, most beekeepers would consider humanely sacrificing the bees to end their suffering. It's a very personal decision—similar to putting down a pet that is dying. Of course, the combs and equipment would be saved, but all of the bees would be sacrificed because they were not productive and would not accept a new queen.

Soapy water kills bees in a few seconds. Mix 1 cup of ordinary liquid dishwashing detergent with 1 gallon of water. Place the solution in a pail and brush or shake the bees into the solution. You can spray bees that are not on combs with this same solution until thoroughly wet, but do not spray the combs with soapy water, as it leaves a harmful residue.

# Using Honey Bees for Pollination

Colony management for pollination is very similar to basic management for honey production. Hobby beekeepers rarely participate in pollinating agricultural crops, but they certainly should know the basics of pollination by honey bees. Neighbors and city officials are easier to deal with when you reveal the true value of bees in our society. Approximately 20 billion dollars worth of crops produced in the United States each year require honey bee pollination. Without honey bees, these crops couldn't be produced. Honey bee pollination provides one-third of our daily food.

Why are honey bees so useful as pollinators? There are many reasons. Honey bees are easy to propagate, relative to any other pollinator species. You can transport hives easily to target crops when and where they are needed. Bees are available throughout the production year. Their long flight range is impressive in that they can reach more than 40,000 acres from their hive location. They are great cross-pollinators, because individual bees forage exclusively on the same species, thereby spreading the correct pollen to cause fertilization. Yet another major advantage is that, as a total population, they forage on a diversity of plant species. All of these attributes in combination account for the honey bees' great contributions to pollination and food production.

Hobby beekeepers provide a valuable pollination service in urban environments where other pollinators, such as many species of wild bees, are rare. Fruit trees—apple,

# CROPS THAT ARE POLLINATED BY HONEY BEES

| Common Name | Product of Polination | | | Common Name | Product of Polination | | |
|---|---|---|---|---|---|---|---|
| | Fruit | Seed | Nut | | Fruit | Seed | Nut |
| Alfalfa | | X | | Currants | X | | |
| Almond | | | X | Eggplant | X | | |
| Apple | X | | | Fennel | | X | |
| Apricot | X | | | Guava | X | | |
| Avocado | X | | | Kiwifruit | X | | |
| Beans | X | X | | Macadamia | | | X |
| Beet | | X | | Mango | X | | |
| Berries | X | | | Mustard | | X | |
| Broccoli | | X | | Okra | X | | |
| Brussell sprouts | | X | | Onion | | X | |
| Buckwheat | | X | | Papaya | X | | |
| Cabbage | | X | | Peach | X | | |
| Cantaloupe | X | | | Pear | X | | |
| Caraway | | X | | Peas | | X | |
| Carrot | | X | | Persimon | X | | |
| Cashew | | | X | Plum | X | | |
| Cauliflower | | X | | Quince | X | | |
| Celery | | X | | Rapeseed | | X | |
| Cherry | X | | | Safflower | | X | |
| Chestnut | | | X | Sainfoin | | X | |
| Chile pepper | X | | | Sesame | | X | |
| Clover | | X | | Soybean | | X | |
| Coconut | | | X | Squash | X | | |
| Coffee | X | | | Strawberry | X | | |
| Coriander | | X | | Sunflower | | X | |
| Cotton | | X | | Turnip | | X | |
| Cucumber | X | | | Watermelon | X | | |

cherry, pear, plum, apricot, and so on—require cross-pollination to produce fruit. Many trees and shrubs need pollination to produce food for birds and wild animals. All berries need bee pollination. Gardens need pollinating, too, to produce melons, squash, and many other crops. Instead of anti-beekeeping ordinances, cities should encourage beekeeping by developing educational programs and commonsense regulations based on real issues, not imaginary and emotional issues. Beekeeping benefits everyone.

## Relocating Hives

Sometimes there is a need to relocate hives in your backyard, perhaps to the other side, if a neighbor complains that your hives are too close to his property for comfort. Moving a hive too far too fast can create problems. Foragers remember their hive location to within inches, even if they have flown miles. If their hive is missing when they return, then they will be confused, flying around and searching for their hive. If they can't find it, they will cluster on some object near the original hive location. To relocate hives within your yard, the hive should be moved short distances—about 10 to 20 feet each day—until the new destination is reached. Bees need a day of flight at each temporary stop along the journey so they can reorient.

If you need to move your hive(s) to a distant location, be sure that the new apiary is at least five miles from your home apiary. Otherwise, many long-distance foragers may return to your home apiary. However, this would not be a problem if another hive remains at your home apiary to accommodate the displaced foragers from the hive that was moved.

A canola field in bloom is awesomely beautiful. Bees pollinate oil crops, too.

Before you move your hive(s) a long distance from home, you should consider the consequences. Is the relocation really worth the effort and expense? The old saying that the grass is greener on the other side of the fence really applies. Moving your hive(s) near a colorful field of flowers can be tempting, but don't underestimate the superiority of foraging opportunities in urban areas. Competition from commercial apiaries in a country setting can reduce honey production per colony. There may be hundreds of hives nearby. You can have more fun and take better care of your hives when they are easily accessible in your own backyard. Furthermore, keeping your hives at home protects them from unknown risks in countryside locations, including pesticides, ants, skunks, vandalism, and theft. However, if you want to produce a specific kind of honey—citrus honey, for example—then you must place your hive(s) by the citrus grove for the duration of the honey flow.

# How to Move Hives

It seems so simple—just load the hives in your vehicle and drive like crazy to the new destination. Not so fast—there are easy and safe ways to move hives, but this requires some preparation. Don't emulate professional beekeepers, who move their hives at night. Bees are very defensive at night—they cling and sting. The best time for a hobby beekeeper to move hives is at daybreak, when there's just enough light to see—and before there is bee flight.

Prepare your hives a day or so before moving them. Be sure that your colonies don't occupy more than two deep chambers. Secure them tightly together with tie-down straps so they won't separate during transport. The most convenient vehicle for moving hives is an open trailer or a truck with an open bed. You should ask a friend to help you load and unload the hives. Just two chambers with a fair amount of honey can weigh well over 100 pounds. Know precisely where you are going, and don't stop to refuel along the way.

The greatest hazard to your colonies during transport is overheating. Bees need ventilation—lots of it—especially during warm weather. Screening the entrance may seem like a good idea. However, when bees are confined by an entrance screen, they get very excited. Their activities quickly cause the internal colony temperature to rise. This stimulates bees to forage for water to cool the hive. These bees, combined with foragers and guard bees, all trying to leave the colony at once and pressed frantically against the screen as they are attracted to light, cause a bee stampede. Within minutes, the entrance screen becomes completely plugged with bees trying to escape the heat, their bodies essentially blocking all ventilation.

In this situation, the colonies can undergo a meltdown—literally—in which the internal heat can get hot enough (around 144 degrees Fahrenheit) to melt the combs. All of the bees die in a meltdown. This potential disaster is easy to prevent. Replace the hive top with a securely attached ventilation screen so that the entire hive top provides functional ventilation. Be sure to spray water on the screen to keep the bees cool. Repeat the water application if the trip is long or whenever bees begin to cluster under the screen. With this upper screen in place, you may now safely cover the hive entrance with screen. The key to moving colonies is to keep moving, never stop for more than a few minutes, and unload promptly upon arrival.

It's risky to transport hives enclosed inside your trunk or vehicle. If the hive "leaks," the bees will be attracted to light and fly to the windows as they try to escape. Bees released inside a vehicle are not a threat unless they distract the driver. Hives moved in an open trailer or truck bed receive better ventilation because of the wind. If you have to move hives without a helper, then a hand truck is useful. Tilt the hive backward. A sideways tilt could shift frame positions and crush bees, including the queen. You aren't obligated to move both chambers at the same time. Just divide them and carry them individually to the vehicle, then stack them and strap them together. Do the reverse at your destination. If a chamber is too heavy, simply transfer several combs to a holding container—maybe an extra empty hive body—and replace them (in the same order) in the chamber at the vehicle. Divide and conquer (and save your back).

**Moving hives to another location is simple. Secure the hive, provide full top ventilation, screen the entrance, spray the top screen with water, and then off you go.**

Use lots of smoke before you release the bees at the new location. Bees tend to be more defensive at the moment of release and for several days after the move. Do not place the hives where flying bees could cause a problem; for example, near roads, pathways, farm animals, field workers' favorite shade trees, or locations where farm implements might accidentally knock them over.

Swarms not captured by beekeepers can develop unwanted colonies in wall cavities.
These wild colonies may be very large. Bees can build combs to fit almost any cavity shape.

# Swarm Prevention

Capturing a swarm is a fun way to start your new colony, but you don't want your own bees to swarm once your colony is established. With few exceptions, swarming is a springtime phenomenon, at least for European honey bees. In the old days, beekeepers actually welcomed swarming as a means of starting new colonies. Now, swarms are considered a problem. First, the loss of foraging bee populations greatly reduces honey production in the colonies that produced the swarms. Secondly, in urban settings, swarms are unwelcome invaders that disturb people in residential backyards, on school property, in public places, and so on. A cloud of flying bees terrifies most people because the people are sure that the bees are going to sting them. (Of course, the opposite is true. As previously mentioned, swarming bees are not a significant sting risk.)

Swarms create major problems when they set up housekeeping in undesirable places. What homeowner wants to discover that thousands of bees are nesting in the walls of their home for an indefinite period of time? Most homeowners dislike the idea of exterminating bees because they are so beneficial. In addition, the expense of extermination by a licensed professional is daunting.

You must use swarm-prevention procedures during the swarm season, a period of approximately six weeks in late spring when colony populations peak. You need to inspect your colonies weekly during this period. The colonies should have two brood chambers. Using a single brood chamber encourages swarming, because the bees are too crowded. Discovering multiple queen cells signals the onset of swarm season as well as the need for swarm-prevention procedures. One commonly recommended method is to destroy all queen cells once a week. This method doesn't work well in practice. It's hard work, and there is a good chance that you may overlook a cell. That one cell will undo all your efforts. Even if you destroy all of the queen cells, the bees' natural urge to swarm is so pervasive that they'll start new cells again. Sometimes they'll swarm even if all of the cells are destroyed.

A better approach to swarm prevention is to mimic natural swarming by suddenly reducing the population of the colony, thereby eliminating the crowded conditions within the colony that stimulate swarm preparations and eventual swarming. One option is to divide the colony. You will need additional hive equipment—an extra top and an extra bottom board. When you first notice multiple queen cells—usually more than six in various stages of development—divide your colony in half to create two single-story colonies, as follows. Put one chamber on the extra bottom board and cover it with the extra top. Now you have a new colony—commonly called a *divide*. You don't need to find the old queen. Be sure that both colonies contain combs with brood, queen cells, honey, and somewhat equal bee populations. Immediately move the new hive to a location at least 2 miles away—perhaps to your buddy beekeeper's apiary—to stabilize the population. If you were to leave the new hive in the same apiary, most of the bees would return to the familiar parent colony location. Place a honey super over a queen excluder on each of the divides to relieve the crowded brood chambers.

## BAD BEE PR

Homeowners shouldn't have to deal with unwanted bee colonies originating from the hives of neighboring beekeepers. The problems that are created when stray swarms infest private homes do not improve the image of bees or beekeepers, and such problems encourage cities to consider ordinances against keeping bees. Hobby beekeepers have a major responsibility to prevent swarming.

Inspect both colonies one week after making the division. You don't have to search for the old queen. She will be in the colony that contains eggs and very young larvae. By this time, the bees should have destroyed the queen cells in the colony containing the old queen because the colony is no longer crowded with bees. In the other colony, you should see queen cells, perhaps capped and still developing or maybe uncapped, indicating successful emergence of a virgin queen. It's best to avoid inspecting this colony for a minimum of four weeks—five would be better. The queen needs this quiet time to mature, mate, and start the new brood nest. This is the time to identify her by marking, tagging, or clipping one wing. If all goes as planned, you will have stopped swarming and simultaneously produced a new and vigorous queen—for free. Now you can move the distant colony back to your home apiary.

At this point, you have another decision to make—do you want to have two colonies or just one? If the answer is one, the solution is simple. Allow both of the divides to produce brood for another three to six weeks; then find and sacrifice the old queen. You can pickle her in rubbing alcohol so you can show your friends what a queen bee looks like or donate her to a local biology teacher. Display a worker and a drone with her for comparison. Now combine the two colonies, thereby creating a very strong, super-productive colony with two brood chambers on the bottom and two honey supers above a queen excluder. This colony won't swarm because the swarm season has passed, and you will have a new, young queen.

If you want two colonies, you have a few options. One is to keep the old queen and not to combine the two divides. Another option is to find and sacrifice the old queen when you first divide the parent colony so that both divides can produce new queens. Still another option would be to replace the old queen with a purchased mated queen.

If you don't want to move your divide to a different location, there is another option: the Demaree swarm-control method. This method was first published by George Demaree in 1884. First, find the queen. Place her—still on the frame of brood where you found her—in the middle of an empty brood chamber positioned on the bottom board. Destroy any queen cells on this frame. Fill out the chamber with empty combs and confine the queen in this chamber with an excluder. Place two honey supers on the excluder and add a second queen excluder on top of the honey supers. Finally, place the original brood chambers above the second excluder. Destroy all queen cells in the upper two brood chambers. Examine the upper two chambers a week later to destroy any queen cells that may have been constructed in the interim period.

This is a typical hive size for a population in late spring—about the time when swarm preparations may be underway. A queen excluder separates the two deep brood chambers on the bottom and the two shallow supers on top.

Rearranging the hive configuration can prevent swarming. Most of the brood is now on top, above a second queen excluder, while the queen is below the lower queen excluder with empty brood combs ready for more eggs.

Enough bees will migrate to the lower brood chamber to support the queen. The two honey supers in the middle provide enough separation between the upper and lower brood chambers to prevent normal queen pheromone distribution. Bees in the upper chamber therefore behave as if there is no queen in the colony; this stops their inclination to swarm. The advantage of this method is that all the bees and brood are retained in one colony and are functional. After the swarm season has passed, you can restore the colony to its original configuration.

A variation of the Demaree method is to substitute a moving screen for the upper queen excluder and provide a small entrance for bees in the isolated chamber to fly rather than destroy the queen cells in the upper chamber. There is a good chance that the virgin queen will mate and provide you with a new young queen up top. If this succeeds, then you can sacrifice the old queen below and reconfigure the hive. Now you have stopped swarming and requeened the colony in one operation. If you're unlucky and don't produce a new queen up top, you can simply reconfigure the colony after swarm season has passed.

## Queen Replacement and Introduction

Queens that are not performing well should be replaced. An entire season's honey production can be lost by the compassionate beekeeper who takes a wait-and-see approach and gives the failing queen more chances to improve her performance. Poor queen performance produces a weak colony that is much more susceptible to diseases, parasites, injury by predators, and robbing by other bee colonies. You have to decide which you like better: nursing old queens in poor health or producing honey. Replacing the failing queen with a young, vigorous queen is easy if she is properly introduced. You have to keep in mind that she is foreign to the colony population and may be attacked and killed. Aggressive bees crowd tightly around her in a ball-shaped cluster and attempt to sting her. Beekeepers refer to this as "balling" the queen.

There are several ways to introduce replacement queens safely. Most hobby beekeepers purchase replacement queens from commercial queen breeders. The replacement queen arrives by mail in a cage that contains several workers to care for her during transit. A compartment on one end is filled with sugar candy that provides food and temporarily blocks the escape opening.

In order for the queen to be accepted, she should be confined inside the shipping cage somewhere near the middle of the brood nest. You do not need to remove the attendant worker bees. Use rubber bands, tape, wires, or any means to attach the cage, but just be sure that the screen side is exposed to bees outside the cage so they can feed her though the screen. During this time she will be fed lavishly, her ovaries will begin to function again, and her pheromone secretions and other odors will begin to match the expectations of the bees in the hive. In order to keep the bees from eating

**Queen bees are routinely mailed to beekeepers. During transport a cage contains several worker bees to nurse the queen and some sugary fondant for food.**

the candy too quickly and releasing the queen prematurely, keep the cover on the candy for the first three or four days. Then remove the cover to expose the candy at the exit hole so the bees outside the cage can eat the candy, thereby releasing her within the next twenty-four hours. If she is released without this prolonged confinement period, she is likely to be rejected and killed by the bees.

Another dependable queen introduction method is a simple push-in cage. You can make this cage from #8-mesh galvanized wire cloth. Place the queen on the comb surface and quickly confine here there by pushing the open back side of the cage deeply into the comb—as near to the foundation as possible—so the workers won't chew around the edge and release her prematurely. Manually release the queen about four to five days later with minimal colony disturbance. Examine the colony about a week after the queen's release. The queen's successful introduction will be indicated by the presence of eggs and young larvae. This is a good time to mark the queen. Marking her after she starts laying eggs is easier because she is heavy with eggs and can't fly. Remove the empty cage and respace the combs properly.

**A new queen must be introduced properly to be accepted. This push-in cage is just one way to confine the queen on the comb during her introduction.**

The advantage of the push-in cage method is that the queen has access to the comb and can start laying eggs earlier. The disadvantage is that you have to transfer the queen from the mailing cage to the push-in cage. Her abdomen shrinks during transit, so she is now able to fly. If she accidentally escapes, you may never see her again. But there is a chance, after her orientation flight is over, that she will return to the release point, where she can then be captured. Any time you must handle a queen that is capable of flight, you should work near a window—indoors or in your vehicle. She would be attracted to the light and could be recaptured quickly on the window pane if she is released accidentally.

Many beekeepers maintain a reserve queen throughout the active season so they can quickly replace a failing queen at any time. Early in the season, establish a small, nucleus colony in a five-frame hive that can be purchased from bee-equipment suppliers. Any time a queen begins to fail, simply sacrifice her, but don't leave her body in the hive. Remove a frame of brood containing the reserve queen and insert it—with adhering bees—into the brood chamber immediately after sacrificing the old queen. The new queen is instantly accepted, and egg laying continues without interruption. Place the displaced frame in the nucleus colony. Chances are good that a new queen will be reared in the nucleus colony. Check one month later to confirm this. If not, the combs and bees can be combined with the requeened colony.

A miniature hive, known as a nucleus hive, is open to show a total of five brood frames. It's ideal for maintaining a backup queen in case of an unexpected failure of a queen heading a production colony.

# Diseases of Bees

Diagnosing health issues in bees may be more challenging than diagnosing problems in humans. Bees can't tell us where they hurt—and taking their blood pressure is a real chore. Their size also complicates the research tasks. The number of honey bee scientists researching bee diseases in the world is far less than 1 percent of the scientists researching human maladies. The good news is that bees don't get cancer and they have no problem whatsoever with athlete's foot.

Sharing beekeeping equipment, especially combs, with another beekeeper greatly increases the chances for spreading infections. Small apiaries of hobby beekeepers receive some protection because of their isolation. Prophylactic treatments are available for some diseases, but most hobby beekeepers choose not to treat their hives because the risk of disease is small. It's probably best to avoid prophylactic treatments in your hives, as there are almost always unknown risks, such as contaminating the combs with chemical residues that may be harmful to developing brood.

Many diseases affect bees. The microorganisms associated with these diseases affect honey bees exclusively and are harmless to humans. There are entire books on the subject of bee pathology. Several of these bee diseases are especially serious. One bacterial species, (*Paenibacillus larvae* ssp. *larvae*), merits special attention because it causes a serious disease known as American foulbrood (AFB). Colonies can become infected when beekeepers unknowingly introduce spores by placing contaminated brood combs into healthy colonies. Spore introduction can also happen when foragers collect honey left behind in wild colonies that died of AFB.

Young larvae up to three days old ingest spores that germinate in their gut and kill them later as older larvae or young pupae. As the dead brood decays, it "melts down" on the lower cell wall, turning a dark brown and eventually drying and adhering tightly to the lower cell wall. As the disease advances, there is a recognizable foul odor in the brood nest. Spores form in the decayed remains, and some are trapped between the cocoon layers that accumulate in the brood cells. These spores are viable for more than forty years in contaminated combs and hive equipment. As bees are refurbishing cells, some spores are released. Once the colony is infected, you should destroy the combs by burning them. Otherwise, the disease can spread from hive to hive, especially when brood combs are interchanged. You can salvage the hive equipment by scorching the interior walls to kill the spores.

Every time you inspect a hive you should examine the pattern of capped brood cells. An excessive number of scattered empty cells in the capped brood area is the

**Symptoms of American foulbrood are scattered cell cappings as well as sunken, discolored, and perforated cappings. Larval and pupal remains lie flat in the cell-bottom wall, eventually producing a brown scale that clings tenaciously.**

first clue that there may be an AFB infection. Then look for sunken, discolored, perforated cappings. Housecleaning bees attempt to clean out the remains and, in the process, they perforate cappings, leaving telltale evidence of AFB. Dip a toothpick into the gooey, brown, decaying mess and withdraw it slowly; the material will "string out" in a smooth, slightly elastic thread.

If you find these symptoms, you should not destroy the colony until a qualified professional confirms your tentative diagnosis. Some states provide the services of professional apiary inspectors who can make field diagnoses of certain bee diseases. To find such inspectors, visit the Web site of Apiary Inspectors of America (www.apiaryinspectors.org). Another option is to search the Internet for "honey bee diagnostic services."

## Parasites of Bees

A large number of internal and external parasites infest bees. One of the most serious parasites is an external mite, *Varroa destructor*. These large mites suck blood from adults and developing brood, especially drone brood. The mites severely stunt developing bees inside the capped cells or cause damage to body parts, especially wings. They also compromise the vitality of adults. In addition to the damage inflicted by direct parasitism, these mites transmit other serious diseases, especially viruses. Severe infestations cause the colony to become unproductive or to die.

Hobby beekeepers must learn to identify and monitor populations of these mites as well as to use contemporary preventive and control measures. Over the years, many miticides have been developed and marketed. However, the mites quickly developed resistance to each chemical. The best long-term solution is for professional bee breeders to develop and maintain bees that are resistant to the mites and to make these bees available to beekeepers.

Hobby beekeepers can keep Varroa mite infestations at acceptable levels by placing a drone comb in the brood nest during the active season to attract the mites. Each time the drone brood becomes capped, place the comb in the freezer overnight, and then return it to the colony. Bees will remove the dead brood and clean the cells, and the queen will lay eggs to start the cycle again.

**This injured drone pupa is parasitized by a Varroa mite.**

Other serious parasites attack honey bees internally. For example, there are several species of *Nosema* parasites—protozoa that destroy the lining of the adult bee gut. Aside from the direct effects of parasitism, these protozoa also have a high potential for spreading viruses and other microorganisms. Accurate diagnosis of *Nosema* by the hobby beekeeper is difficult. Few hobbyists become skilled as bee gastroenterologists. Get some professional help on this one.

## Insect Pests

One of the worst insect pests is the wax moth (*Galleria mellonella*). The female moth lays eggs at night in the cracks and crevices outside and inside the hive. Tiny wax-moth larvae tunnel through the combs—eating as they go—and lining the tunnels with silk that affords some protection from the bees, which would cast them from the colony. As the moth larvae grow larger, they are more exposed and vulnerable, and populous bee colonies defend themselves very well. Maintaining populous colonies is the beekeeper's best strategy for controlling wax-moth larvae in the hive. If the colony population declines for any reason—disease, a failing queen, and so on—the moth larvae may win the battle. Within a few weeks, especially in warmer

**Silk-lined tunnels indicate the beginning of a wax-moth infestation. Without immediate treatment, the comb could be totally destroyed within a few weeks.**

climates, the moths will have destroyed all the combs, which will be transformed into a tangled mess of webs littered with larval fecal pellets.

Stored combs are very vulnerable to wax-moth damage, especially in warmer climes. In northern states, combs can be stored outdoors in unheated structures from late fall to early spring. In warmer areas, the combs can be frozen (at less than 10 degrees Fahrenheit for 24 hours) to kill all stages of the wax moth. If these combs are immediately sealed in plastic bags with *absolutely no openings*, they will remain free of moths even when stored at room temperature. Do not return frozen combs to supers for storage unless the supers were also frozen. Otherwise, eggs in the supers will hatch and the combs will be destroyed. Very cold combs are quite brittle, so handle them carefully.

Another insect pest is the small hive beetle (*Aethina tumida*). The best defense against this pest is to maintain populous colonies. It's also important to protect stored combs. Combs can be frozen, as previously directed for wax-moth treatment, to kill all stages of

**When the comb is completely destroyed, it will be mass of webbing and larval feces. The adult moth is shown here.**

the small hive beetle. This is a relatively new pest. Strategies for control are changing rapidly, so it is important to seek current information.

In some parts of the country, ants can be a serious pest. They invade the hive and steal the goodies. In extreme infestations, bee colonies may slowly decline in population until they are no longer productive. Using pesticides, especially airborne sprays that could be "inhaled" by the colony via the entrance, near the hive is extremely risky. If the hive is on a platform with legs, ants can be deterred using various sticky substances, repellants, and pesticide-treated barriers to prevent their access.

# Animal Pests

Skunks can be annoying pests, even in some urban environments. They scratch around hive entrances at night, causing bees to emerge. Skunks, which are apparently immune to stings, feast on bees. In one incident, a skunk was found to have sixty-five stings inside its mouth, and its stomach was stuffed with bees. Frequent nocturnal disturbances by skunks seem to make the bees more nervous—more sensitive to disturbances—thereby increasing your chances of being stung when you are opening colonies that have been bothered in this way. Controlling skunk attacks on hives is important. A fence around your hives should work well. For other options, search the Internet for "skunk control in the apiary."

Mice may damage the unprotected lower combs during the winter. Cover the entrance with #3-mesh galvanized hardware cloth to prevent mice from entering while still permitting the passage of bees.

Bears can destroy entire apiaries overnight. If your apiary is located in bear country, you could lose all of your colonies in one nocturnal episode. Bears love honey and relish bee brood. Electric fences—ideally installed *before* the apiary is discovered—seem to work well. For more options, search the Internet for "bear control in the apiary."

Each geographic area has a different pest profile. The list of creatures that are pests to bees is very long and ever-expanding, as new pests are accidentally imported from other areas of the world. The best way to obtain current information is to read beekeeping trade journals, attend bee-club meetings and national conferences, and search the Internet frequently.

# Pesticide Threats

Pesticides are a serious threat for professional beekeepers who must situate their apiaries in agricultural settings near large acreages of monocultured crops and orchards that require a high degree of insect control. Fortunately, most hives owned by hobby beekeepers are in urban environments where pesticide risks are minimal. Colonies tolerate small amounts of pesticides in gardens, home orchards, and other restricted areas—although a few bees may fall victim. Hobby beekeepers are sometimes their own worst enemies when they inappropriately apply pesticides near their hives or stored beekeeping equipment.

# CHAPTER 10

# Honey & Other Hive Products

Bees consume most of the honey they make. Honey is primarily food for them and secondarily a treat for us because they produce more than they require for their sustenance, which is 200 pounds per colony annually. The extra honey—anything over 200 pounds—is known as "surplus" honey because it can be harvested without jeopardizing colony survival. Production of surplus honey varies dramatically in different parts of the country. Hobby beekeepers usually expect to produce around 100 pounds of honey per hive per year. Consider how you can make other people happy by surprising them with honey gifts.

# Honey: The Ultimate Natural Treat

Honey has been prized as one of the most delectable natural foods since primitive man discovered that wild bee nests could be raided to collect this wonderful sweet stuff. By definition, honey is made by honey bees from nectar secreted by flowers. Because honey varies considerably in color, flavor, and moisture content, it is always unique in its qualities. Harvesting honey is very different now from when wild nests were the only source. Modern hives make the harvesting job relatively easy and negate the need to destroy the combs in the process. Honey is still the same as it was thousands of years ago: delicious, natural, and full of energy.

## Composition of Honey

Honey contains an average of about 17 percent water. Sugars comprise the remainder; these include around 38 percent fructose (also known as *levulose*), 31 percent glucose (also known as *dextrose*), 7 percent maltose, and 7 percent of other miscellaneous sugars, including 1 percent sucrose (ordinary table sugar). Other ingredients—including acids, complex natural chemicals that account for flower fragrance, enzymes, traces of minerals, and pollen grains—are present in trace amounts. These trace components do not significantly contribute to human nutrition. (No one wants to eat 15 pounds of honey daily to satisfy the minimum daily requirement for a trace element.)

**Hot waffles and fresh honey make breakfast a delicious feast, especially if you produced the honey in your back yard.**

Honey is essentially a watery solution of sugars. To a great extent, these sugars affect the physical properties of honey: tendency to granulate, viscosity, and the absorption of water when exposed to moist air. Honey is very acidic, averaging a pH of around 3.9. Scientists have researched details of honey composition extensively—far more than you need or want to know unless you are a chemist as well as a beekeeper.

## Granulation of Honey

Most honeys granulate during storage after extended periods of time in containers. Sometimes honey granulates while still sealed in the comb. The basic reason honey granulates is that the bees have dissolved more sugar in the solution—a process called *super saturation*—than it can hold during storage. The tendency to granulate is determined primarily by the concentration of glucose. Excess glucose forms crystals of *glucose hydrate* that aggregate in a lattice in the honey.

Microscopic "seed crystals" incorporated during extraction start the process of granulation. At completion, honey may become solid or have a suspension of crystals in the liquid. The optimum temperature to promote granulation is around 55 degrees Fahrenheit. Lower or higher temperatures slow the process. At very low temperatures, around 0 degrees Fahrenheit or below, granulation essentially stops, providing a convenient method for very long-term storage in a jar or in the comb.

Granulated honey in a glass container is easy to liquefy in a microwave oven, but be careful. Heat it in thirty-second intervals—stopping and stirring. Monitor the temperature so you don't have to heat more than necessary to achieve liquefaction. Heating honey to a temperature of 160 degrees Fahrenheit for thirty minutes dissolves all of the seed crystals, which greatly extends shelf life. However, high temperatures also cause chemical changes that some purists consider to be heat damage. The nutritional quality of heated honey remains essentially the same. However, heat may change delicate flavors and cause the color to darken.

When honey granulates naturally, the crystals tend to be large and feel like particles of sand as you eat the honey. By controlling the granulation process, you can produce a very smooth, creamy consistency that is pleasant to eat and use as a spread. When you're eating a piece of toast, you may want to slather it with honey that won't drip, similar to peanut butter. You can induce granulation by adding previously granulated honey to liquid honey. Make a mixture at room temperature that is about 5 to 10 percent granulated honey and the rest fresh liquid honey. Store it at about 55 degrees Fahrenheit. Do not warm the granulated "starter" honey to make it soft and easily manageable. Heating could destroy the crystals that are needed to cause granulation. If all goes well, you can achieve a very smooth, non-grainy spread. Mix some butter or other goodies—limited only by your imagination—with the "creamed" honey to create a tasty treat. It's nice to have a choice of eating honey as a liquid or a spread. Some kinds of honey won't granulate, however. A good example is honey produced from the nectar of tupelo trees in Florida and Georgia. Honeys that won't granulate naturally are regarded as premium honeys and priced accordingly.

## Fermentation of Honey

Honey in its natural state has a high concentration of sugar-tolerant yeast species. These yeasts won't grow and cause fermentation unless the moisture in honey is around 20 percent or higher. Fermentation during honey storage produces a foul-tasting product that is neither mead (see Glossary) nor honey but something in between. You can minimize the risk of fermentation during storage by harvesting fully cured honey from capped cells with moisture content less than around 18 percent water. You can easily measure moisture levels in honey with a little gizmo called a *refractometer*. Take a sample of your honey to a beekeeper who owns one, and you'll see how one drop of honey tells all—providing your sampling technique is valid. The surface of honey exposed to the air can be high in moisture—not representative of the whole lot—so sample beneath the surface. Granulation releases additional moisture. This additional water can push stored honey with borderline high moisture content over the edge, initiating unwanted fermentation.

## Hygroscopic Properties of Honey

Honey is hygroscopic, meaning that it readily absorbs moisture from the air. That is why gourmet cooks include honey in the recipes of baked goods—to prevent drying and crumbling. The texture of the baked good is more pleasing and, as a bonus, you may taste some subtle honey flavor, depending upon the quantity of honey in the recipe. When you mail a batch of holiday cookies made with honey to Aunt Gertrude, they have a better chance of arriving as cookies instead of crumbs.

Before harvest, bees control the humidity inside the hive and protect honey from hygroscopic dilution. Once you remove the supers of honey from the hives, hygroscopic action begins. Honey produced in very dry climates is low in moisture and extremely thick—very difficult to dispense for consumption. Exposing combs in the supers to moist air in a warm room for a day or so just prior to extraction may increase the water content and make it easier to pour. However, in humid climates there is the option, and sometimes the need, to remove some moisture before extraction. You can arrange supers of combs for active ventilation in a very dry, warm environment for a day or two. This removes some moisture even if the honey is capped. A dehumidifier is also handy for this purpose.

## Harvesting Honey from the Hive

In their neverending search for an easier way, beekeepers have invented many ingenious methods and devices—similar to the search for a better mouse trap—to remove bees from the combs. Some commercial beekeepers use repellent chemicals to drive bees downward into the brood chambers before removing the supers. Hobby beekeepers have better options.

A quick way to harvest honey from a small number of hives is to shake and brush the combs free of bees. If you have a leaf blower, just blow the bees off. Ideally, do this during midday, when foraging bees are most active. Smoke the colony well, anticipating maximum disturbance. If you have bee gloves, this is a good time to wear them. Remove the honey supers and set them aside on a supporting structure. Remove the queen excluder. Place an empty chamber (no frames) on top of the hive. Remove each frame of honey, smoke it thoroughly, and then shake the bees into the open hive top. Shake hard, as they won't fall free with weak shaking motions. Use a bee brush to dislodge the residual bees downward into the open chamber. Transfer the bee-free frame into an empty receiving chamber. Prevent the entry of robber bees by covering this chamber with a screen closure.

The quickest way to harvest combs of honey is to shake and brush bees from the combs.

Repeat the process for all frames. Now repeat the process for the next honey super. When you're done with that hive, the empty chamber on top of the brood chamber will have lots of bees clinging to the interior walls. Smoke heavily and brush most of them down into the brood chambers. Replace the hive cover, and you're done harvesting honey from this hive.

I've mentioned that deep honey supers can weigh around 80 pounds. Do not risk getting "beekeeper's back." Transfer a few frames at a time to a smaller container that you can carry safely. Move the harvested honey indoors (but not into the kitchen if you want to preserve your marriage). There are always a few loose bees that fly up and toward lights or windows.

An alternate method is to insert a partition equipped with a "bee escape" between the brood chambers and the honey supers. Several models of bee escapes are available commercially. Bees trapped above the partition escape downward through the one-way bee escape. If the weather is warm enough, virtually all of the bees will abandon the honey supers to join their cohorts in the brood chambers below; this usually requires twenty-four to forty-eight hours. There are two important requirements for this method. First, there must be no brood in the honey supers. Bees are reluctant to abandon brood. Secondly, there must be no openings large enough to admit a bee in the honey supers above the partition. Otherwise, foraging bees in the area—from your hives or from other sources—may discover the unprotected honey and show up by the thousands to steal it. During the resulting feeding frenzy, they may also invade the brood chambers, rapidly overwhelming the guard bees. Opposing populations will be fighting and stinging each other, and the disturbance will likely cause the death of the queen. All the honey may be taken away within a surprisingly short time. In addition, the bees may become very defensive during the melee, stinging people and animals in the immediate area.

## Harvesting the Winter Honey Stores

You can increase your honey production greatly by not overwintering your hives and instead harvesting the winter stores—perhaps an additional 40 to 60 pounds of honey. Confine the queen in a small screen cage in the brood chamber during the fall to terminate brood-rearing. About twenty-two days later, all of the brood will have emerged, and the honey flow will have ended. Harvest all the honey, including the winter stores in the upper brood chamber, and shake the bees into the lower brood chamber. Take your hive to your buddy's apiary, and then shake and brush all of the bees from your hive so they can join other colonies. Store your combs safely and start a new colony the next spring from a swarm or package of bees. You could freeze several frames of honey as a convenient way to feed your new colony. By not overwintering your colony you have (a) increased harvestable honey, (b) avoided colony mortality during the winter as well as the problems that develop when a colony dies, (c) prevented swarms the following spring, (d) ensured that you will start each season with a new queen, and (e) protected your hive equipment from exposure to the elements during the winter. As a bonus, your buddy will appreciate the boost in population for overwintering his hives and may agree to provide you with bees to start your colony the next spring.

## Extracting Honey

Congratulations! You are now reaping the rewards of hobby beekeeping. Extracting honey is fun. Some beekeepers use the honey-extraction process as an excuse for a celebration, inviting neighbors and kids down the street to share the excitement and receive samples of fresh honey. A typical extracting setup (in a typical home's

garage) includes equipment for uncapping honeycomb cells, removing honey from the cells, and dispensing filtered honey into storage containers. First, the combs must be warm enough to reduce the viscosity of honey. Cold honey is just too thick to extract efficiently. It must flow easily from the comb cells

**A typical honey extracting setup in a hobby beekeeper's garage. Wax cappings are removed with a hot knife, combs are placed in the honey extractor (centrifuge), and honey filters as it is drained into the plastic container.**

by centrifugal force provided by a *honey extractor*, a centrifuge designed to hold comb frames. You may delay buying one until you've had bees long enough to establish your hobby. In some areas, you can rent a honey extractor from a nearby bee-supply store. A better solution may be to make a deal with your beekeeper mentor, who is already set up for honey extracting. You'll need to see how it's done before you can do it yourself. The extraction process can be challenging to a beginner working alone—even with good instructions in hand.

First, you have to remove or perforate the cell cappings to expose the honey. You can easily remove cappings if you used either eight or nine frames (instead of ten) in the honey supers, resulting in thicker combs with cappings that protrude enough to provide easy access with an uncapping knife. A heated electric knife works well for removing cappings, and a cappings scratcher is also handy, especially when the cappings are difficult to reach with the knife. The cappings fall into a receiving chamber on a drain surface. Tubs with perforated bottoms for this purpose are available in beekeeping-supply catalogs. Some beekeepers improvise by attaching a queen excluder (the parallel-wire type) to the bottom of an empty super.

**Beeswax cell cappings fall away, revealing beautiful fresh honey. The honey-soaked cappings drain and the wax can be salvaged.**

To uncap a frame, hold one end in a vertical position with your left hand and temporarily support the other end on a sharp point, such as an upward protruding nail, as if you are going to rotate the comb on end. Support the frame end on the nail and proceed to "saw" upward with the knife, cutting just deep enough to remove the cappings. The slab of cappings, wet with honey, slides downward

Honey is flung from the eight spinning combs to empty them so that they can be returned to the hives for refilling.

over the hot knife, warming the honey in the process, and falls. You need to lean the comb toward the knife so the slab of honey will fall free without sticking to the exposed honey in the cells. It's important to note that as you are pressing upward and cutting the cappings, you should anticipate that the knife will occasionally fling upward. Keep those fingers hidden behind the end bar at the top of the frame, or you'll be sorry. You can cut downward if this seems easier for you.

The warm, wet cappings will start draining immediately. Chop them into smaller pieces now and then to facilitate draining. After each comb has been uncapped on both sides, it's ready to be transferred into the honey extractor.

Many hobby beekeepers choose a reversible extractor. In this design, the centrifugal force flings honey out of the cells on the outer side of the comb. Honey on the opposite side cannot be extracted until the frame is reversed (rotated 180 degrees). The name of this type of honey extractor is derived from the need to reverse the frame position twice to get all of the honey out of both sides of the comb. There are three steps. In the first step, you spin the combs around very slowly to empty approximately three-quarters of the honey from cells facing outward toward the tank wall. If the speed is too great, the centrifugal force of honey on the opposite side of the comb will seriously damage the comb, especially if the comb was made from beeswax foundation. In the second step, you reverse the frames so that the side with full cells now faces outward. Use a fast speed in the second step to completely empty the cells. The third step is to reverse the combs once more to complete the extraction of the first side with another fast spin. Extracted honey impacts

A two-frame reversible honey extractor is perfect for a hobby beekeeper. The two frame baskets swing in either direction so honey is emptied from one side at a time.

the side walls of the centrifuge/tank, then puddles temporarily at the bottom reservoir. Periodically open the valve at the tank bottom to drain the honey into a 5-gallon plastic food-grade pail that you can use for long-term storage.

Some beekeepers prefer a radial extractor, in which centrifugal force is applied equally on both sides of the comb toward the frame's top bar. Start the centrifuge slowly, and honey flows freely from all of the cells simultaneously as the speed increases.

Some honey is allowed to stay as a ballast to help stabilize the extractor, but the rotating frames will hit it if the level gets too high. At this stage, the honey contains bits of wax. It's good to strain the honey as it flows from the valve and falls into the collection pail. An item that you really need is a stainless-steel double-sieve strainer that rests on top of the pail and receives honey flowing from the extractor-tank gate valve. This type of strainer has a large mesh strainer on top and a fine mesh strainer

**This simple strainer works well for small batches of honey, but there may be small wax fragments in the honey. A double-sieve strainer is better.**

on the bottom. When the strainers get plugged with wax fragments, sugar crystals, and so on, you must clean them; but never use hot water, or the beeswax chips will melt and permanently plug the fine mesh. Simply immerse the filters in warm water and swish them with circular motions to dissolve the sugar crystals. Backflush with sprayed water to remove other foreign matter.

During the extraction process, the cappings drain slowly. Honey accumulates below in a "cappings tub" or similar container fitted with a gate valve. Later, you'll filter honey drained from the cappings through the same double sieve strainer. Allow the cappings to drain in a warm room overnight to salvage most of the honey. You can wash cappings in warm water to remove the residual honey so the wax can be salvaged for other projects, such as making candles.

## EXTRACTOR STABILITY

Extractors never seem to be balanced perfectly despite your best efforts to match the weight of opposing combs. You must securely anchor the extractor to a heavy, solid base to enhance stability.

When you're done extracting, you'll have a stack of empty combs that are wet with honey. Don't store them in this condition. Place them back on your hive for cleanup by the bees. To do this, wait until late afternoon, use lots of smoke, remove the hive lid, replace the queen excluder on top of the brood chambers, and place one or two supers of wet combs on top. Replace the hive lid. It's best to put wet combs on all hives in the apiary at the same time. Otherwise, the "have not" colonies may start to rob the "haves." If there is a honey flow in progress, these combs are immediately

**Combs wet with honey should not be exposed outdoors for bees to "clean up." There can be serious problems, such as stimulating robbing behavior and spreading microorganisms, including American foulbrood spores.**

available for a refill. Otherwise, the bees will quickly salvage the honey and store it in the brood chamber below. In a day or so, remove the honey supers containing dry, empty combs and store them.

Remember to protect the combs from pests during storage. If you should decide to discard any items—combs, cappings, and so on—that are wet with honey, be sure to place them in a container that prevents access by bees.

## Processing and Storing Honey

Honey usually granulates at some point during storage. How are you going to liquefy a 5-gallon block of hard, crystallized honey? Place the container in a very warm environment for twenty-four to forty-eight hours until it is partially liquefied. Stir and then pour it into a *bottling bucket*, a heavy-gauge 5-gallon bucket fitted with a scissors-gate valve at the bottom that is used to dispense the honey into smaller containers. Dispense the honey into pint or quart jars that can be heated in a microwave until the honey is completely liquefied. Alternatively, you can place the jars in a pan of water heated near the boiling point. Stir now and then, and monitor the heating progress until the honey becomes liquefied. You can monitor the temperature with a thermometer to prevent overheating; honey should not be heated above 160 degrees Fahrenheit or kept hot for more than thirty minutes.

You can avoid the hassle of granulated bulk honey by simply dispensing the honey into small containers soon after extracting, before it has the opportunity to granulate. Transfer the freshly extracted and filtered honey to the bottling bucket and fill the containers while the honey is fresh. Honey weighs about 12 pounds per gallon. If lifting 60 pounds of honey in a 5-gallon container is difficult, then fill the bottling bucket to a level that will be safe for you to handle.

Honey that is to be dispensed at the table should not be stored in the refrigerator. Cold honey is too thick to dispense easily. Either store containers of honey at room temperature for frequent use or place them in the freezer for long-term preservation—even for years.

## Eating Honey in the Comb

Honey is at its best when it is sealed in the comb, commonly referred to as *comb honey*. When honey is extracted, it flies through the air in tiny droplets and spreads thinly on the interior walls of the extractor tank. This exposure to air causes some flavor loss. You can enjoy the full flavor of honey by eating bite-size pieces of freshly constructed honeycomb. You can also chew beeswax like chewing gum. Swallowing beeswax is optional and doesn't benefit you in any way because it is indigestible.

**Honey directly from the comb is the ultimate natural food treat. All the flavor is there to enjoy. Slice little squares of comb and spear with a toothpick. You can eat the wax, too, or chew it like chewing gum.**

Here is a simple way to produce a small amount of comb honey. Remove a comb frame from a honey super. Replace it with a frame with the comb removed except for a 1-inch strip at the top to encourage bees to build new comb in the frame. At the beginning of an expected honey flow, place the frame in the center of the honey super immediately above the queen excluder. Bees will fill in the frame with new comb. The honey is ready to be harvested when capped. Cut the comb into small squares, seal the squares in plastic bags or plastic cut-comb boxes available from bee-equipment suppliers, and then preserve the squares indefinitely in the freezer.

## Honey Varietals

Beekeepers sometimes relocate their hives to capture honey flows from specific floral sources. For example, citrus honey can only be produced by hives located near citrus groves. Honey from specific floral sources tends to have a distinctive flavor and color—a floral signature—especially if there is insignificant incoming nectar from other floral sources at that particular time. Varietal honeys are prized for their unique flavors. The fragrance and flavor of citrus honey is very similar year after year; it reminds you of the heavenly fragrance of citrus trees in bloom.

It's almost impossible to produce honey that is 100 percent from a particular plant species. Foraging populations in a given hive typically forage on multiple sources at the same time. These honeys get mixed together—if not in the hive, then at the time of harvest. Honey varietals are classified according to the predominant flavor rather than by the percentage—always an unknown value—of a certain floral source in the mixture. The best honey varietals are made by simply placing empty honey supers on the hive at the beginning of the honey flow from the plant species of interest and harvesting the filled combs immediately after the honey flow—before honey flows from other plant species begin.

# Other Hive Products

In some parts of the world, bee brood and adult bees are considered delicacies themselves. Chocolate-coated adult bees are a novel treat. Another product of beekeeping is royal jelly, the food of the developing queen-bee larvae. Royal jelly is frequently a minor component in cosmetics. Some ingredients of royal jelly are being investigated for potential medical uses, especially as anti-tumor treatments and antibiotics for microorganisms that cause human diseases. Many unfounded medical and nutritional claims have been made for royal jelly as well as for other hive products. Future research will continue to define the benefits to human health and nutrition. Be cautious and do not use yourself as a guinea pig. Bees produce many amazing biologically active agents, and as these agents are discovered and properly researched, they are likely to benefit mankind in the future.

**Two workers are chowing down (see the extended tongues) on royal jelly in this young queen cell with the cell wall partially removed. The larva is visible.**

## Pollen as Human Food

Many people believe that bee-collected pollen has special nutritional qualities that promote human health. Athletes on winning teams consume pollen in the belief that it gives them greater energy and endurance. However, athletes on losing teams do the same. Pollen is difficult for humans to digest, so an undetermined percentage passes through the human alimentary canal without being digested. Proteins, vitamins, minerals, and other nutrients are definitely present in pollen, but they are trapped inside the microscopic pollen grains, which have tough walls. More research is needed to determine the digestibility of pollen and its nutritional value for humans.

If you want to try it, you can collect fresh pollen from hives in your apiary. Several commercially manufactured pollen traps are available. These traps attach to the entrance so that pollen-laden bees returning from foraging are obliged to pass through small openings to enter the hive. Pollen loads on the hind legs protrude outward just far enough to be dislodged as bees squeeze through the trap openings. Pellets of pollen fall through a screened area and collect in a tray or container below. You should collect pollen daily,

**Pollen colors vary according to plant sources. These pellets were collected from bees entering the hive entrance.**

remove stray particles of foreign materials, and package it in sealed bags for storage in the freezer. Don't trap pollen for extended periods, though, or you may deprive the colony of enough pollen for proper nutrition. You should be able to see many cells of pollen in the brood chamber at all times.

Some people have reported allergic responses caused by eating pollen. Be cautious. Try very small amounts at first to test for a possible allergic response and be prepared in case you have a reaction.

## Propolis

Some of the chemical compounds in propolis are biologically active and may have important medical uses. Research indicates the possibility of using chemicals found in propolis for many applications, including treatment of burns, bacteria, fungi, tumors, and even dental plaque. Scientists need to conduct clinical studies of propolis to document the potential benefits and hazards of its medicinal uses. Such studies are very challenging because the composition of propolis is highly variable. No two samples are precisely alike. However, there is a reasonable expectation that biologically active components of propolis can eventually be isolated, identified, synthesized, and subjected to controlled clinical studies. Until that time, medicinal use of propolis is risky.

There may be great medical benefits or there may be nasty toxic compounds, even carcinogens, in the mixture of chemicals found in natural propolis. Hobby beekeepers should avoid conducting experiments with propolis, especially for consumption or dermal applications. Leave propolis research to professionals.

Perhaps the most interesting application of propolis is not medical. It has to do with musical instruments. The famous Stradivarius instruments—especially stringed instruments made between 1698 and 1725—reportedly were treated by Antonio

Stradivari with a special varnish that contained propolis. Maybe propolis enhanced the sound quality of these wonderful instruments?

## Bee Venom

When you get stung, you may actually be receiving free health benefits. Bee venom is biologically active in many ways. Once again, it is difficult for medical researchers to obtain clinical evidence because the composition of venom is variable, even when applied directly from the stinger. Nevertheless, scientists are conducting extensive research in attempts to discover new ways to use bee venom for improving human health. As previously mentioned, there is extensive empirical evidence that bee-venom treatment (apitherapy) has been successfully used for treating some kinds of arthritis and rheumatism. Scientists have developed procedures for collecting large quantities of venom for research and medical applications by literally "milking" venom from the bees with an apparatus that applies a mild electrical shock. The bees do not lose their stingers and are not hurt by this treatment.

## Beeswax

The primary sources of beeswax are cell cappings removed during honey extraction or recovered from brood combs that become obsolete with age. Beeswax can be melted, filtered, and refined to some extent by various processes. It has literally dozens of applications, especially for making candles, cosmetics, and honeycomb foundation. Beeswax candles are superior to other wax candles because they burn brighter and longer, they do not bend, and they burn cleaner than other candles. They are also used symbolically in religious ceremonies around the world.

Save wax cappings and melt them to make a cake of pure beeswax. This is the first step in preparing beeswax for many applications, such as making candles. Be careful: beeswax is extremely flammable if it gets too hot, and you

**A wick can be rolled up tightly inside a colorful beeswax sheet to make a simple, beautiful candle. Search the Internet for detailed instructions.**

should never heat it over an open flame. Candle-making kits with instructions are available from most bee-supply companies and some craft stores.

**Save wax from cappings to make candles. Many molds are commercially available.**

## Mead

If you carefully control the fermentation of honey in a winemaking process, the result is a tasty product known as mead (honey wine)—probably the first alcoholic beverage enjoyed by man. Mead is a delightful alcoholic drink made by fermenting honey with water, yeast, and traces of nutrients needed by yeasts for optimum growth. It ranges in alcoholic content from a mild ale to a strong wine and may be still, carbonated, or sparkling. Mead is frequently semisweet or sweet, but it can also be dry. Mead-making is a companion hobby practiced by many hobby beekeepers. Another favorite alcoholic beverage is honey beer, brewed in the same way that other beers are brewed, except that honey is substituted for sugar. The flavor varies significantly according to the kind of honey that you use.

**Making mead is simple and this delightful drink can be sweet or dry, depending upon your recipe.**

## Honeydew

Some insect species, such as aphids and scales, feed on plant sap. They secrete a sweet liquid that bees collect as it if were nectar. It is processed like nectar and stored like honey. It can't be called honey because, by definition, honey is made from nectar secreted by flowers. So it is called *honeydew* and is frequently regarded as a delicacy. It contains complex sugars and has a flavor similar to molasses.

# CHAPTER 11

# Fun Things to Do with Bees

The fun things you can do with bees are limited only by your imagination. You can make bees construct unusual comb shapes that are quite artistic, train them to collect food from the palm of your hand or anywhere else, record their sounds outside and inside the hive (you'll be surprised to hear their little sounds inside the hive), and photograph them in the hive and on flowers. Take a long walk and search for plants where bees are foraging. Don't forget to take the kids with you so they can learn, too. Observe the behavior of individual bees as they collect nectar and pollen. It's different for each plant species. Check out housecleaning behavior by dropping small bits of colored paper inside the hive and watching housecleaner bees carry them out to their garbage dump. Use your imagination and have fun. Please just avoid risky experiments or demonstrations that make non-beekeepers nervous or upset.

# Capturing a Bee in a Cage

If you can catch a bee, there are lots of fun things you can do, such as educate others about bees. First, you'll need a cage. Cut a piece of galvanized screen wire cloth (#8 mesh is perfect) with tin snips to make a piece that measures approximately 3¼ x 5 inches. Roll the long dimension around a broom handle or a pipe with a slightly larger diameter than a broom handle until the edges overlap about half an inch. While in this tightly rolled position, wrap a fine wire (or piece of fishing line) near each end so the cage holds its shape when removed. Close the open ends using a cork or a piece of heavy-duty aluminum foil pressed against the ends and crimped into place.

**This homemade bee cage is easy to make and very useful if you need to confine a few bees for some reason. Be sure to handle it by the ends to prevent an accidental sting.**

Remember this: when a bee is disturbed or confined, it attempts to walk or fly upward, toward light. Just point the closed end up, and the bee will obediently move upward long enough for you to close the open end at the bottom of the cage.

The easiest method to capture the bee is to place the open end of your cage over a bee that is resting on a comb surface. When you refine your bee-catching skills you can even catch a bee on a flower while she is foraging. Approach upwind so she can't smell you, don't make fast movements, time your catch maneuver while she is occupied with sucking up nectar, and place the cage over her. Disturbing a honey bee on the flower will never, ever be a sting risk. If you botch the catch attempt, she will just fly away.

# Low-Temperature "Anesthetization"

You may want to make a bee motionless for some reason—maybe to paint her thorax for identification or to remove a pollen load. Just put the caged bee into the refrigerator—not the freezer—until she becomes motionless, which will take maybe ten to twenty minutes. Take her out, dump her from the cage, and do your thing. She will recover completely after a few minutes. Do not bother chilling the queen. She will never sting you.

## Sting Demonstration

If you want to see how a stinger works, catch a worker bee in a small cage, chill her as previously directed, carefully pick her up with tweezers, and press the tip of her abdomen to a piece of suede leather or a piece of animal skin. After the protruding stinger catches by the barbed tip, gently pull the bee away and watch the stinger pump for several minutes. Use a magnifying glass to see the details of this amazing sight. Be sure to smell the stinger when it is first deposited. The bananalike odor is the alarm pheromone odor that excites other bees to sting.

Do not forget to immediately dispose of the stinger donor humanely by dropping her into a glass of soapy water or putting her in a freezer. Don't be squeamish about sacrificing a bee for educational or research purposes. If you drive a car or walk outdoors, you are sacrificing many insects, including bees, by squishing them underfoot or colliding with them—just examine your front bumper or windshield. That's just how life is.

## Capturing "Baby" Bees

The best way to introduce kids to bees—and bees to kids—is to show them freshly emerged adult bees up close. You can call them "baby" bees—kids love that. Select a brood frame in which bees are emerging from their cells. Shake and brush *all* bees off the comb and into the brood nest. Place the comb inside a warm room. Within a few minutes, several bees should emerge. Gently pick them up and place them into a transparent plastic cup that has a thin layer of lubricant—petroleum jelly or vegetable oil—on the inside wall near the top to prevent their escape. (They can't fly at this age, but they can walk out of the cup.) When you're done, return the bees to the brood chamber.

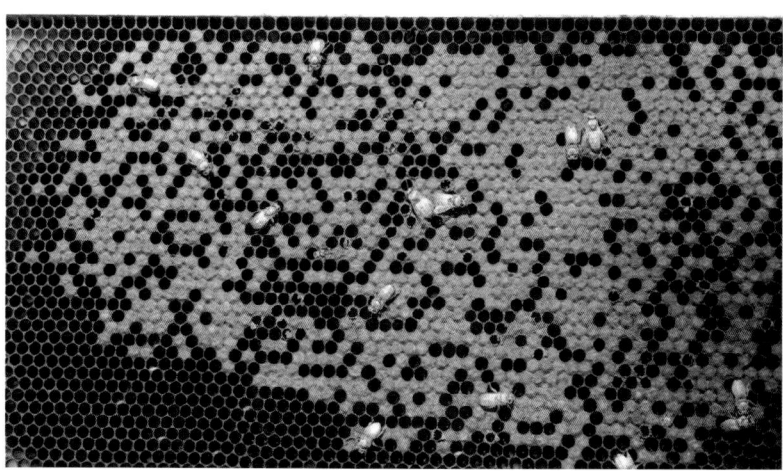

**When a mature brood comb, with all bees removed, is held in an incubator for several hours, many "baby" bees emerge.**

**Returning foraging bees remember and return to the exact location where they departed from their hive. If you move the hive, even if only a few feet, the bees hover at its original location, momentarily confused.**

## Bee Memory Demonstration

Bees returning from foraging flights remember their precise hive location. To demonstrate this phenomenal memory, gently move your hive (with your friend's help) about 5 to 10 feet sideways on a day when there is active bee flight at the hive entrance. Now move away from the hive and watch from the side. Returning bees come to a screeching halt in mid-air in front of the original hive entrance location and hover there as if bewildered. After two or three minutes, quickly and gently return the hive to its original position and stand back again. There will be a mad rush of bees going into the hive entrance in the familiar location. Now they're "happy." If you were to leave the hive out of position a little longer, some returning bees would find the entrance and start fanning and releasing orientation pheromones from their scent glands, which are visible near the tips of their abdomens.

## Observation Hives

There is no better way to learn about bee activities inside the colony than by mounting a glass-walled observation hive inside your home. Observation hives are available from bee-supply companies. In addition, there are do-it-yourself construction plans on the Internet. Be sure that you use the single-comb-width style so you can see both sides of all combs at all times; otherwise, the queen may not be visible. You can also connect the hive to the outdoors via a visible runway that is 1 to 3 feet long and covered with glass or plastic to permit observations of bee traffic at the entrance.

The hive should contain a minimum of two deep frames—four are better. You can use a queen excluder to confine the queen to the lower brood comb(s) as a means of preventing overpopulation.

## Leashed Bees

Just as you can take your dog for a walk, you can take a bee for a flight—something you might do to add a little excitement to an educational lecture at a school or just for fun. Glue a fine thread on top of a drone near the posterior end of his abdomen. (A tiny droplet of almost any fast-drying glue works fine.) When you're done, the thread easily pulls free from the waxy cuticle without harming the drone. Release the drone back into his hive.

## Selling your Honey

Bottle your honey in containers that bear your private label, which you can easily make using one of a variety of software

**Producing more honey than you and your friends can eat? Try a "for sale" sign and you may be surprised by its effect.**

programs and some printable sticker sheets. Some bee-supply companies print labels for honey containers. You'll have no difficulty selling your honey for premium prices. Just advertise that it is fresh, homemade, and locally produced.

# GLOSSARY

**Africanized bees**—hybrids of the African honey bee and various European honey bees

**American foulbrood**—a bacterial disease that kills bee brood

**Antenna**—a thin sensory organ on the head

**Apiary**—a place where beehives are kept and bees are raised for their honey

**Apis mellifera**—the scientific name for the honey bee

**Apitherapy**—the medical use of bee products, especially bee venom

**Bee bread**—stored pollen containing honey and secretions

**Bee smoker**—a device used to create smoke to control bees

**Bee space**—a space of approximately 5/16 inch that bees leave between combs

**Beeswax**—a wax secreted by bees and used for honeycomb construction

**Colony**—the bee population living within the hive

**Compound eyes**—eyes made of multiple light-sensitive parts

**Demaree swarm control**—a method to prevent swarming by rearranging the hive chambers, placing brood on top and confining the queen below in a chamber with empty brood comb cells

**Drone**—a male honey bee

**Epinephrine**—a synthetic form of adrenaline used to treat anaphylaxis

**Foundation**—a beeswax sheet that guides honeycomb construction

**Hive**—a structure used to shelter a honey bee nest

**Hive tool**—a thin metal device used to open hives

**Honeycomb**—hexagonal beeswax cells constructed together for honey and brood storage (frequently referred to as comb when not filled with honey)

**Honey extractor**—a centrifuge used to remove honey from honeycomb

**Honey flow**—a time when there is abundant nectar available for collection

**Honey stomach**—an expandable organ used by bees when transporting nectar and water

**Hygroscopic**—capable of absorbing moisture from the air

**Lancet**—part of the stinger with serrations that anchor the stinger in flesh

**Mandibles**—structures used for biting and shaping wax into honeycomb

**Mead**—an alcoholic beverage made from fermented honey

**Miticide**—a substance that kills mites

**Nectar**—a sweet, watery liquid that attracts pollinators to plants

**Nosema**—a unicellular intestinal parasite

**Nucleus hive**—a small hive used for temporary bee colonies

**Ocelli**—simple, light-sensitive eyes on top of the head

**Ommatidia**—light-sensing units fused together to form the compound eye

**Package bees**—living bees shipped in screen cages

**Parthenogenesis**—development of an egg without fertilization

**Pheromone**—a secretion that affects behavior and development

**Pollen**—the male reproductive cells of flowering plants

**Pollen basket**—a concave outer surface of the bee's hind leg, used to hold pollen

**Pollen substitute**—a mixture of nutrients that mimic pollen

**Pollination**—transfer of pollen from male to female part of flower

**Proboscis**—mouthparts used for sucking liquids

**Propolis**—a resinous substance from buds used as caulking and cement by bees

**Proventriculus**—a valve in the alimentary canal that controls the one-way entrance of liquid food from the honey stomach into the ventriculus where digestion takes place

**Queen**—the only fully-developed, egg-laying female in the bee colony

**Queen cell**—a special vertically oriented cell used for producing a queen

**Queen excluder**—a device used to confine the queen to part of the hive

**Rectum**—an expandable intestinal organ used for storage of feces

**Refractometer**—an instrument used to measure the percentage of moisture in honey

**Robber bee**—a foraging bee that collects honey and nectar from honeycomb

**Skep**—a basket-shape bee hive made of woven straw

**Small hive beetle**—a small beetle from South Africa that eats hive products, especially pollen

**Stinger**—a sharp organ used to inject venom

**Swarm**—a large population of bees flying or clustered on a surface

**Tarsus**—the foot structure of an insect (the plural is tarsi)

**Ventriculus**—the part of the alimentary canal, immediately posterior to the proventriculus, where digestion takes place

**Wax glands**—glands on the underside of the abdomen that secrete beeswax

**Wax moth**—a species of moth that produces larvae that consume honeycomb

**Worker bees**—sterile females that perform most of the activities in a bee colony

# RESOURCES

## Useful Web Sites

### International Beekeeping Information

**APIMONDIA Foundation, www.apimondia.org**

APIMONDIA promotes scientific, ecological, social, and economic apicultural development in all countries. It promotes cooperation of beekeepers' associations, scientific bodies, and individuals involved in apiculture worldwide, and it organizes international congresses on all aspects of beekeeping.

**International Bee Research Association, www.ibra.org.uk**

The world's longest-established apicultural research publishers, IBRA promotes the value of bees by providing information on bee science and beekeeping worldwide. IBRA is a not-for-profit organization founded in 1949.

### National Beekeeping Organizations

**American Beekeeping Federation, www.abfnet.org**

The Web site provides current beekeeping news in the United States. The American Beekeeping Federation also sponsors an annual North American Beekeeping Conference and Tradeshow.

**American Honey Producers Association, www.americanhoneyproducers.org**

This Web site focuses on the needs of beekeepers specializing in honey production. It provides useful links to beekeeping information on the Internet. The association also sponsors an annual conference.

**Apiary Inspectors of America, www.apiaryinspectors.org**

The Web site provides information for apiary inspections and lists by state the names and offices of entomologists, apiculturists, and apiary inspectors who provide apicultural services and information.

**Eastern Apicultural Society, www.easternapiculture.org**

The Eastern Apicultural Society is a nonprofit educational beekeeping organization. The Web site lists excellent internet links for beekeeping information. The society meets annually in various locations in the eastern United States.

**Heartland Apicultural Society, www.heartlandbees.com**

The Heartland Apicultural Society is a non-profit educational beekeeping organization. The society meets annually in various locations in the central United States.

Western Apicultural Society, www.groups.ucanr.org/WAS/

The Western Apicultural Society is a nonprofit educational beekeeping organization. The society meets annually in various locations in the western United States.

## Beekeeping Trade Journals

*American Bee Journal*, www.americanbeejournal.com

The *American Bee Journal* is a beekeeping trade journal published since 1861. It includes information on current beekeeping issues, help for hobby beekeepers, popular and research articles, and ads for beekeeping supplies and services. The Web site offers a free sample of a digital journal issue.

*Bee Culture*, www.beeculture.com

*Bee Culture* is a beekeeping trade journal published continuously since 1873. Its issues feature articles on current beekeeping topics and ads for beekeeping supplies and services. The journal comes in both paper and digital editions, and the publisher will send a free sample copy on request.

## Other General-Information Sources

Bee Source, www.beesource.com

The Web site features a very active online beekeeping community forum to promote exchange of information between beekeepers. The site also has general information about beekeeping resources.

Cooperative Extension System, http://www.extension.org/bee%20health

The Web site provides an interactive learning experience and a comprehensive treatment of bee health from the smartest land-grant-university minds across America.

Mid-Atlantic Apiculture Research and Extension Consortium, http://maarec.cas.psu.edu

MAAREC is a regional group focused on addressing the pest-management issue in the mid-Atlantic region. The company has excellent general beekeeping information for hobby beekeepers.

University of Georgia Managed Pollinator CAP (Coordinated Agricultural Project), www.beeccdcap.uga.edu

This site presents information from a nationally coordinated team of experts with proven credentials in extension, genomics, pathology, toxicology, management, pollination, and bee behavior. The project addresses issues affecting the diverse populations of managed bee pollinators across the United States.

# INDEX

# PHOTO CREDITS

# ABOUT THE AUTHOR

Dr. Gary started hobby beekeeping at the age of 15 in Florida. At 26, he earned a PhD in Apiculture at Cornell University. His knowledge is based upon diverse bee-keeping experiences during a 60-year career. He wore many hats: hobby beekeeper, commercial beekeeper, deputy apiary inspector in New York, honey bee research scientist and entomology professor at the University of California, Davis (32 years), adult-beekeeping-education teacher, professional entertainer (bees and music), and professional bee wrangler for 18 Hollywood movies, 6 commercials, and over 70 TV productions. He is the author of more than 100 publications, including scientific papers, book chapters, and popular articles in beekeeping trade journals.